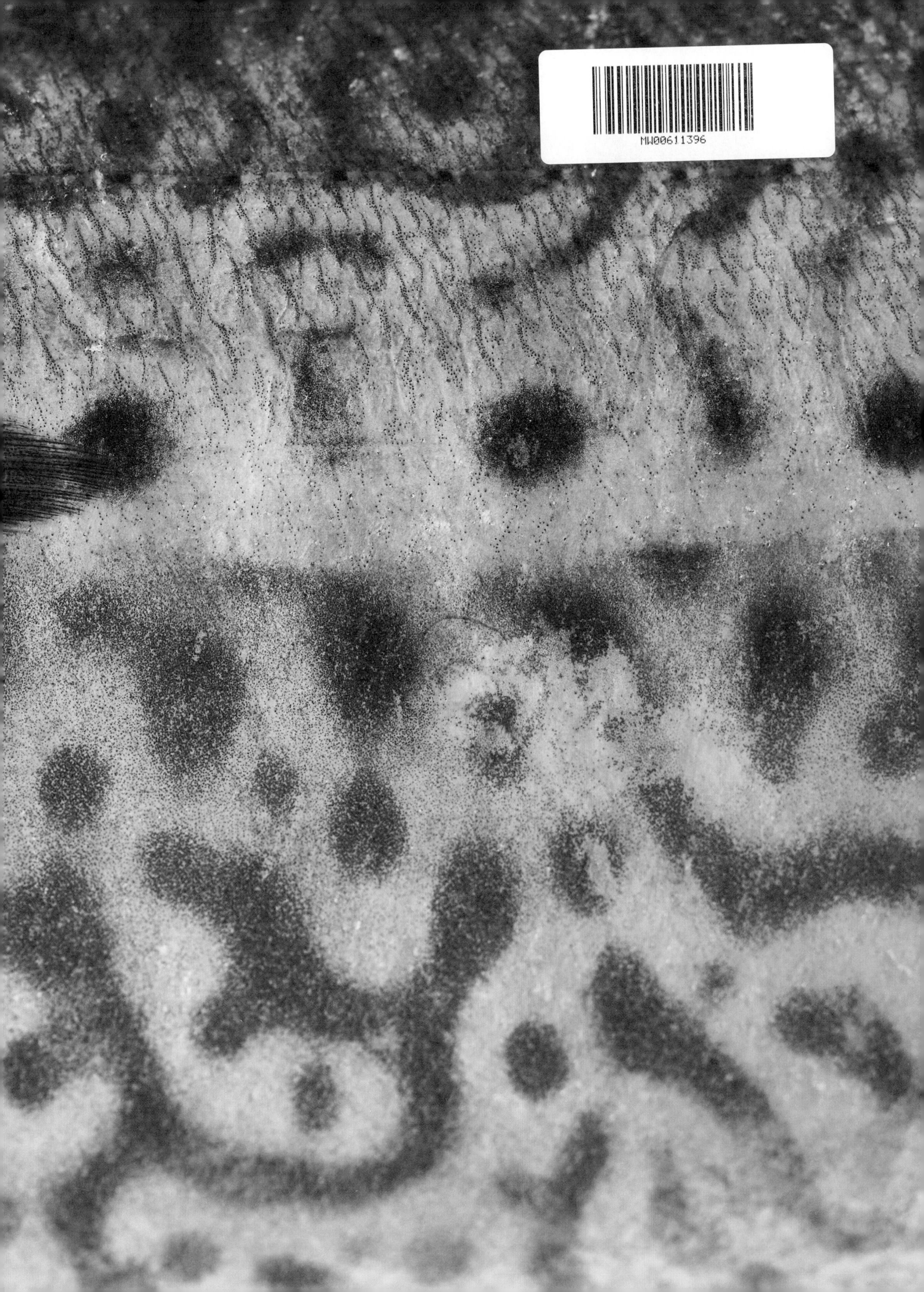
MW00611396

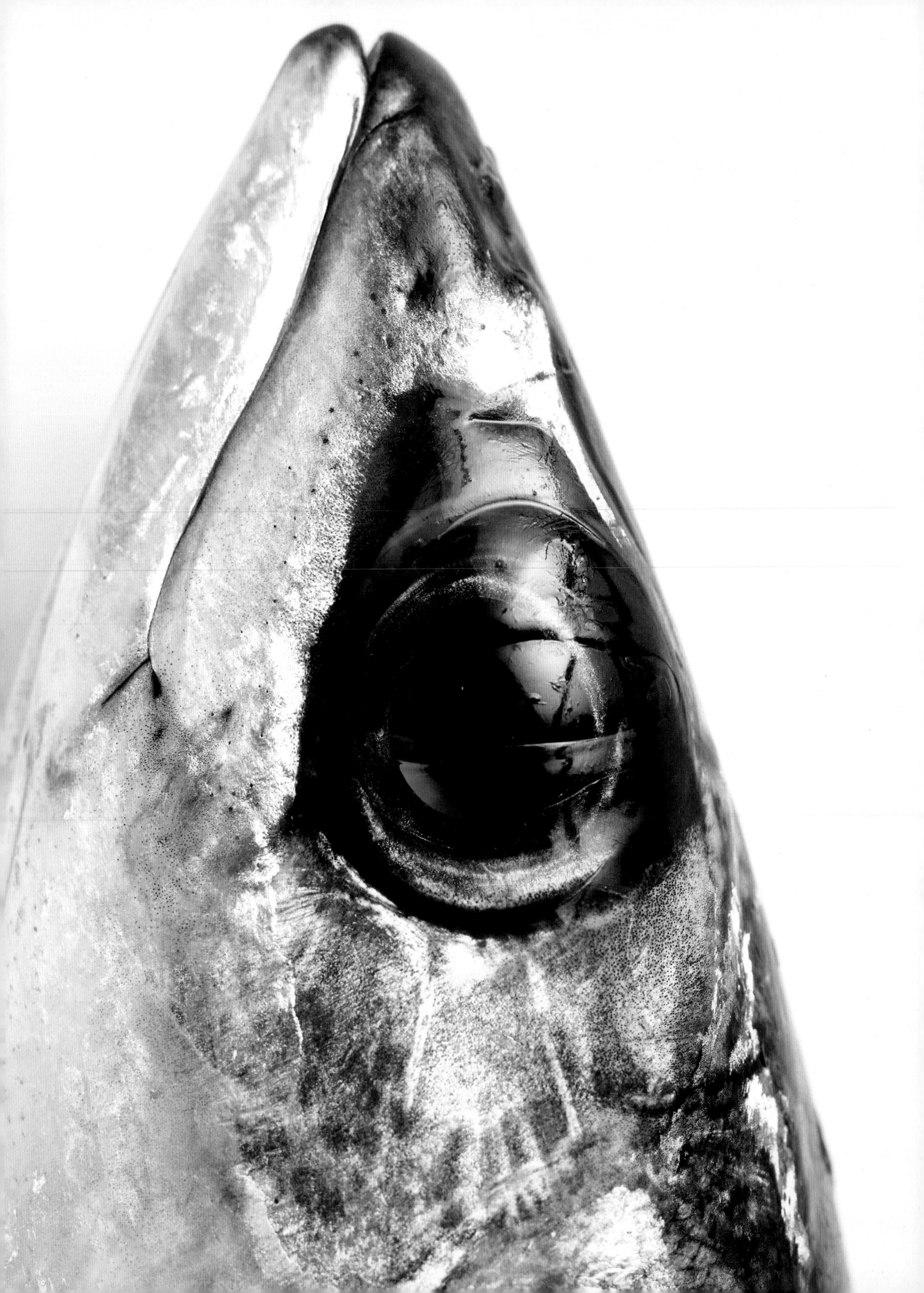

FISH BUTCHERY

Josh Niland

Photography by Rob Palmer

Hardie Grant
BOOKS

It is estimated that 50 per cent
of the world's fish that is caught is wasted.

Of the 50 per cent that we do work with, 50 per cent of this is overlooked by
the Western world, which overwhelmingly prefers to consume only the fillet.

The responsibility for this issue lies first with the industry.

This book intends to disrupt, inspire, challenge
and hopefully encourage the next generation.

Thinking of this as a problem for individuals to solve
is hopelessly out of date.

This book is as much about intentionality as it is sustainability.

CONTENTS

INTRODUCTION (OR THE MARKET FOR LEMONS)

When buying a used car, we have all become accustomed to thinking that it may be a lemon. (We had this experience when we bought our first refrigerated van for Fish Butchery, but that's a story for another day.) If you assume you might be buying a lemon – and have no knowledge of how cars work – then this places a cap on how much you are willing to spend. This phenomenon discourages owners of well-maintained quality used cars to sell them as they won't receive a fair price. In turn, the market fills with more and more lemons.

And so it is with fish.

In every fish shop I have ever seen, fish is washed and stored in direct contact with ice. As a result, it has a very short shelf life, smells 'fishy' and sticks to the pan.

So why should you assume that the fish at our Fish Butchery are any different? Why would you pay double the price for our fish?

The difference is that our fish have been meticulously sourced from excellent local fishers, brain spiked, dry-handled and dry-aged. But how is the average punter to know that?

And so, in fish we have another market for lemons – and a race to the bottom. Consumers know little of the subject and fall back on price as the easiest variable to compare.

However, there is a bright side. The cost of fish is considerably less than that of a car. I know that once you commit to a well-handled fish that has been expertly butchered or transformed into an exceptional product then you will taste the difference and return for more.

Never before has there been such attention on sustainability. Through challenging the status quo and communicating transparently with the industry and its consumers – and, of course, delivering a delicious product – I believe we can drive out the bad, replace it with the good and overcome the market for lemons in the fish industry.

WHAT IS A FISH BUTCHERY?

Before we get started, I need to talk a little about what a fish butchery is, why it's needed and how ours came to be.

The history of meat butchery and meat-based preparations has long been a catalyst for my creativity when trying to reimagine missed opportunities with a fish, and this history also underlines my thinking around the creation of our first Fish Butchery in Paddington.

The meat industry is not without its own flaws, but there would be no butcher in the world who would only strip the loins off a pig and then throw the rest away as waste. And yet this is essentially what we do every day all around the world with fish. Fillets are taken, the rest is discarded and more fish are caught to sustain the demand.

Instead, I believe that when a fish arrives to market, like an animal to a butcher, there are disciplined practices that we need to follow. These practices will vary from fish to fish depending on its size and species. In the creation of our fish butchery, my team and I have developed and refined a number of both basic and complex techniques to encourage desirability and value from the entire fish. These techniques range from butchery cuts to dry-ageing, charcuterie and a list of value-added products that can be made from the whole fish.

It's important to note that we have not created this as a niche concept for a privileged 1 per cent seeking out the extremes of modern gastronomy. This is about creating a workable system for the industry that allows families and individuals to conveniently access approachable fish dishes that extend the use of one fish further, whether that be in the form of fish sausages, burger patties, fish fingers, fishcakes, pâté, mortadella, bacon and more.

By recontextualising fish in this way, we can bring desirability to otherwise overlooked species and parts of a fish.

THE STORY SO FAR

First things first: how did we get to the point of creating Fish Butchery?

Within 18 months of opening my fish restaurant Saint Peter, my team and I had developed a vast array of techniques, dishes and relationships with fishers all over Australia. Given the footprint of the restaurant hadn't expanded and more guests were wanting to experience the work we were doing, my wife, Julie, and I felt it was necessary to open another business.

Not just another restaurant, but a space that allowed for greater infrastructure, a larger team and a connection to our Paddington community that extended beyond a restaurant experience in pursuit of greater standards.

What also played into the decision was our need to diversify where our fish were going. If a single fish arrives on day one, what were the steps we would take to achieve the maximum return? It may be that we would need to produce a terrine, a ham, pâté from its liver, centre-cut fillets, bone-in chops, consommé from the bones, salted and smoked hearts and spleens. But not all these products could be bottlenecked onto one single Saint Peter menu.

If we were going to continue to be a restaurant that only served fish then we would need to create our own fish shop.

A fish shop to me – when I was a kid, at least – was always cold, smelly and wet. So that's where we first put our attention, and we set about changing these three areas to create a more comfortable setting that brought more desirability to fish.

Temperature is critical when handling fish, and we needed to ensure that the space would be kept well cooled while we processed it. However, for the customer, we didn't want water on cold concrete floors, or the traditional ice display covered in fillets of fish that are freshly washed every 10 minutes. That frequency of washing and the inconsistencies of temperature that fish are subject to on an open ice display also contribute to the offensive aromas that you smell even before walking into a market or fish store. By critically thinking about the storage, display and our methods of cutting, we developed a fish shop that found a solution for all three of these issues.

Fish Butchery Paddington was hatched in a former hair salon just a few doors up from Saint Peter on Sydney's Oxford Street. This was by no means the perfect location for a fish shop (albeit a fish butchery), however it was just doors away from the restaurant and had been a cafe many decades ago. We were fortunate, as we were at Saint Peter, to already have a grease trap, exhaust and pre-existing approvals, which meant the wheels could turn quickly and we could get started.

Aesthetically, Julie and I had always been blown away by the orientation of the space at Lune Croissanterie in Melbourne. Unfortunately, the space that we had found in Paddington wasn't as generous, however it still allowed us to be inspired by Lune's ability to bring visibility and transparency to its craft and ingredients.

When we opened Fish Butchery Paddington, the first objective was to display the fish we were going to sell as a single slice or whole fish. These were presented in a static refrigerated glass box at the end of our processing table that would keep the fish chilled from –1 to 0°C (30°F). By doing this, our customers would be able to see a desirable portion of each fish and then select which species and exactly how many portions or grams of each they'd like.

The theory was that we would hold the whole fish on the bone in our coolroom up until the point of being purchased and take it off the bone as it was sold. This would ensure a longer shelf life and a far superior product, as the flesh of the fish wouldn't be exposed to oxygen or light. Furthermore, each customer could ask for exactly what they wanted: a butterflied fish ready for the grill, a crumbed fish ready for pan-frying, or even a boned-out fish, trussed and ready to roast.

However, the failure of this static glass box was that people would come for a look and think that the scarcity of the fish display suggested this was all we had left. As it went so against the conventions of a traditional fish shop's ice trough full of fillets, customers were also confused by what exactly it was that we sold. Beyond attracting guests who merely wanted to grab a photo for their Instagram or come for a giggle at the fish shop with only a dozen portions of fish on display, it was extremely hard to convey the service we were providing.

It was a challenging opening that required a lot of rethinking about how we could make things simpler for people to understand while emphasising our celebration of quality instead of quantity. Communication with visitors to our butchery during this time was of paramount importance so that they understood what they could do, what we could do for them and what was available.

Without hastily shifting the goalposts on the retail display or our level of service to those who did come to buy fish in the early months of trade, the team and I went to work on what the space was primarily intended for – providing Saint Peter with a higher standard of fish that was both more diverse and used a greater percentage of one single fish.

Without hastily shifting the goalposts on the retail display or our level of service to those who did come to buy fish in the early months of trade, the team and I went to work on what the space was primarily intended for – providing Saint Peter with a higher standard of fish that was both more diverse and used a greater percentage of one single fish.

Naturally, as Saint Peter became busier and internal and external expectations grew, we started to see a number of our restaurant customers and locals begin to use Fish Butchery for fish and chips on weekends, or to purchase the occasional fillet of fish for their dinner during their work weeks. Christmas, Mother's Day and Easter were all significant days of trade that helped Fish Butchery solidify itself into people's weekly routines.

The Covid lockdown of 2020 meant that Fish Butchery needed to support Saint Peter as a business while it was closed to the public. During this time we introduced Mr Niland at Home, which would provide meal kits to our local community and those who wanted to experience Saint Peter in their own spaces. At the same time, we reorientated the space at Saint Peter into one long counter for half the amount of guests, a decision made to deliver greater focus and creative opportunity to the restaurant.

The service of creating meal kits for our customers was obviously a very challenging time, however it was also extremely satisfying. We noticed an increase in demand for both retail and takeaway at Fish Butchery during the 18 months that Covid controlled our lives. The team at Fish Butchery was doing a brilliant job and we felt it had the capacity to take on another business to maximise the efficiency of our labour, so, in 2021, Julie and I made the decision to open Charcoal Fish in Rose Bay. The mission of Charcoal Fish was to provide a selection of accessible fish options that allowed guests young and old to choose from excellent grilled or rotisserie-cooked fish paired with beautiful salads and vegetables that were all on display upon entry to the store.

The first month of trade at Charcoal Fish was like nothing I'd ever experienced before. Every day we served more and more guests. It was wonderful to be cooking again! At the end of each week we would discover we had cooked a tonne of fish and another tonne of potatoes for chips.

Having literally been in the trenches cooking for that whole time, I had failed to go and see the team at Fish Butchery. After those initial weeks of opening, I returned to Paddington and was confronted with a team that had been stretched to their limits. The coolroom ceiling had started to cave in due to the weight of the Murray cod that we had been hanging to serve at Charcoal Fish, and the freezer and coolrooms were all full to the brim with the offcuts, offal and sundries.

Saint Peter hadn't even reopened at this point, and we needed Fish Butchery to be preparing and processing fish for both venues. Although not wanting to open a fourth venue, Julie and I decided that to make the work that we were doing at Fish Butchery sustainable – and not just ethically and economically but for our team as well – we would need to open a larger production space. Within 12 weeks, the new Fish Butchery opened at Waterloo, about a 15-minute drive from Saint Peter. This was one of the biggest undertakings we had faced by far. The initial discussions were that the space would be part production, part retail, part takeaway and part home for our fish butchery masterclasses, and this is roughly what has transpired.

TCHERY
FISH BUTC
BUTCHERY

THE FUTURE

I've outlined the genesis of Fish Butchery in detail here for a reason: at some point in time – through success or consumer demand – all businesses face a decision about whether to grow and expand while maintaining and further advancing their core beliefs and best practices. Or, alternatively, deciding to sacrifice quality for quantity or convenience over craft, resulting in (perhaps) better efficiency and greater profit. But at what cost?

Running any kind of fish business, big or small, is incredibly challenging and continues to become even more difficult with the rising cost of labour and goods along with an ever-changing climate.

But if we turn up today and attempt to raise the standard just ever so slightly then we will see change.

Perhaps when we see the entire fish for its limitless potential both as a food resource and for alternate applications in the world around us, the problem won't seem so impossible.

It is those who interact with fish on a daily basis – on the water, in the market, in the kitchen and, ultimately, on the table – who control whether change can really happen.

HOW TO USE THIS BOOK

This book is designed to show the reader – and the industry at large – the vast range of possibilities that exist when that fish is treated correctly, from those initial moments of capture and transport through to being butchered and processed and, finally, making an appearance on the plate. Split into three sections that reflect this journey – Catch, Cut and Craft – the opportunities for adding value to fish at each stage are examined in detail, with step-by-step breakdowns of the various cuts we employ at Fish Butchery and the recipes involved in transforming the building blocks into desirable finished products.

This book is not a call to action for every home cook to make their own fish liver pâté or go about whipping up a quick batch of fish sausages. Instead, I see it primarily as a stimulus for the industry – a whole-fish solution to an existing problem that will help everyone reach a higher return from one single fish.

I do imagine that many who read these pages will be creatively inspired or motivated to implement some of these new methods and practices. I must mention, though, that to cherrypick and waste in order to achieve a singular outcome will put us in no better position than the one we are in right now.

CATCH

This section aims to shed some light on how humans have historically worked with fish. It ranges from examining the role of a 'fishmonger' and the need for us to distinguish between this and what I suggest we call a 'fish butcher' (as the two do very different types of work from the point of view of a fisher and their decision-making) while also discussing critical issues involved in the landing, storing and processing of fish once caught, including dry-ageing, rigor mortis and the role that water can play in the degradation of landed fish stocks. My hope here is to remove the onus from the consumer around the quality and diversity of the fish being sold and place this weight squarely on the shoulders of the industry instead.

I'm very aware that to truly change the culture and cut through stereotypical ways of thinking about fish will require a tremendous amount of skilled labour when looking through a global lens. This section attempts to identify some of the many issues and fractures that exist within the seafood-handling industry and offer some suggestions about how we can bring positive solutions to the ways we interact with our waters in the future without completely starving the world of fish.

It's a beginning.

MONGER VS BUTCHER

I see this book as a scaffold for a curriculum that sees fish butchery as a profession.

But what about the fishmonger? Simply put, a fishmonger is someone who deals and trades in fish. The role of that individual currently extends to procurement, logistics, communication with customers and suppliers, cutting fish for display or for wholesale clients and creating efficiency and profitability.

The role of a modern fishmonger in 2023 needs to continue to execute the above, however the cutting, fabrication and storage of the whole fish need to be relinquished to a skilled fish butcher who innately knows the conversions and yields of any single fish.

Like a butcher of land-based animals, there is an understood and educated theory that is practised to achieve the full outcome for a single animal.

I'm absolutely certain that there is an ethically conscious mindset by the butcher when beginning to break down an animal, however I feel what's more consciously thought about before picking up a knife is the monetary component of the work.

Can you imagine the gross amounts of lost revenue and food waste there would be if a butcher didn't know anything more than to simply cut the primary muscles off a cow or pig and discard the rest? Or if we had no desire to eat the legs from a chicken or the tough sinewy flesh from the tail of a cow?

A butcher of land-based animals focuses on a much smaller paddock of species, with cattle, goats, sheep, pigs and poultry most synonymous with our domestic tables around the world. However, there are thousands of fish species each with their own unique anatomical compositions. Rather than thinking we need to understand every one, let's at least attempt to create some frameworks around more commonly seen fish.

So when the fish arrives at market, what is the current standard?

Staff employed by the fishmongers are instructed to strip the fish of its scales and offal and sell it as a whole piece or, more often than not, remove all the scales, head, gills, offal and the majority of the bones in readiness for it to be sold as fillets.

Efficiencies and convenience result in the use of water throughout processing to keep stations free of debris, blood and scales. The use of large coarse scalers or, worse, a lottery barrel of abrasive surfaces remove the scales. After each step of the cleaning process, water is introduced to ensure speed is maintained and 'hygiene' is upheld. In the end, we either have fillets or we have a whole 'cleaned' fish. If we're lucky, there'll be a loose bucket or separate section where bones, heads, collars and sometimes livers or roe can be purchased.

There are so many issues raised by this process that I'll unpack, but the biggest one is the neglect of being intentional about how to achieve the greatest yield from one single fish. The solution to the crisis we face around the growing percentage of fish wasted globally is *not* to stop consumption, given there are over a billion people who rely on fish as their main source of protein, but rather to see that when handled intelligently and intentionally, one fish can represent the value of two. This responsibility lies with the industry, not the customer. And the solution isn't simple – to cut the head, offal and bones off a fish and then throw them onto ice hoping to sell them is a flawed concept. Even a centre-cut fillet of salmon minus the bones and skin will still strike fear into the most accomplished of home cooks. More is needed, and that's where a fish butcher comes into the conversation.

THE MONGER

A NOTE FROM TONY WEARNE, FISHMONGER

My role as a fishmonger has changed tremendously over the last 10 years. Although it might just be rosy retrospection, it did seem like simpler times back then. There always appeared to be a relative abundance of fish around – there was good fish and there was bad fish, there was cheap fish and there was expensive fish. The aim of the game was to get the best fish at the best price so you could sell it on to your customers and turn a profit.

It was when I'd first started in the industry two decades ago that the idea of sustainability and ethics really started to become a focus. I think the consumer began to want to be more informed about the food they ate (or at least be appeased that what they were consuming didn't have a negative footprint). And at the same time there was a big push from the government through legislation, licensing and enforcing commercial fishing quotas. This impacted the range of species being caught or targeted by fishers, as well as which ones were desirable to the customer. While some wild catches that utilised certain catch methods became niche, fish farming gained momentum. Aquaculture that could tick all these boxes became popular menu items for restaurants. Now my percentage of farmed fish to wild fish is almost fifty-fifty, whereas in the past aquaculture made up only a small percentage of the volume that we sold.

There does seem to be less wild-caught fish around. Whether this is due to the absence of fish or the result of quota restrictions and other external factors is something a fisher is best to answer. There has also been a noticeable loss of seasonality in seafood, perhaps because of climate change. In the past you could almost predict what would be at the markets the next day and what fish would be available at certain times of the year. Lately, shifting currents and unusual weather patterns have blurred the lines of seasonality to the point where I can barely guess what fish might be available through different times of the year.

Another big change that I've noticed is the number of whole fish we sell to our restaurant customers as opposed to fillets. Whether this is a result of the 'nose-to-tail' philosophy gaining popularity or the impact of social media trends on the industry, or something entirely different, is hard for me to say. Nowadays I see my role as not so much buying huge quantities of seafood, filleting it up and selling it but more about sourcing niche products, developing working relationships with fishers and suppliers, and educating customers.

I feel as though the industry as a whole has cleaned itself up. There is now very little 'black market' fish around and a strong onus on the correct labelling of species and country of origin. Fishmongers in general have become more transparent in their dealings and have far more accountability and responsibility than in the past.

One thing that hasn't changed is the value of good relationships. Good operators will always have a quality product for a fair price. In very recent times this has become even more integral to running a successful business. When there is no abundance of produce, we must rely on those relationships we've built to ensure a consistent supply of quality seafood.

THE BUTCHER

A NOTE FROM DARREN O'ROURKE, HEAD BUTCHER AT VIC'S MEATS

There was a time when butchers respected their craft, the animals they plied their trade on and the land from which they came. It was a time when an animal was given a long and cared-for life, harvested and consumed with respect and used in its entirety. The industrialised food systems and increasingly gluttonous consumption of meat has, over the years, led us down a path to a new and alarming way of thinking that has sadly become the norm. Encouragingly, though, the tide is turning, even if we've largely forgotten the ideals, techniques and ethics of conscious consumption that our grandparents lived their lives by willingly and happily, not because they knew no better but, in fact, because they *knew* better.

My path to butchery came via a series of crossroads. A series of fortunate and life-changing events. A meal of grilled sardines, anchovies and fish stew in the northern Spanish town of Oviedo on a deserted beach that, at 25 years of age, made me think of food as something not just simply to sustain life but as one of its most powerful and inspiring pleasures. Ten years in the kitchen with the last few under chef Alex Herbert set some wheels in motion that haven't stopped moving since. Long before it was the done thing to know a farmer or buy from the bush, Alex was sourcing meat, fish and dairy direct from farmers. It was second nature to her and, subconsciously, it became second nature to me too. But if the kitchen lit a fire in my belly for food, acclaimed butcher Victor Churchill and the people I learned from there fuelled that fire and continued to nurture it for these last 14 years. The chef on a 'gap year' as a butcher who had only ever broken the odd lamb after service became fascinated with butchery. The techniques, the terminology, the muscles – everything consumed me and filled me with more questions than I had answers.

Butchery for me starts with the land, the farmers – who are its custodians – and the animals in partnership with one another. Without this holy trinity there would be no butchery. For me it's just unthinkable to not have a fascination and a respect for these foundations. Understanding the land and people sets the scene for how the butcher plays their part; it's just not good enough to disregard all facets of butchery and focus solely on the lifeless chunk of flesh on the block in front of us. I believe all butchers should have a desire to better understand how that piece of flesh came to be on that block, and then the same fascination and interest in the next step – the cooking. Knowing what sets a braising cut apart from a grilling cut, or where the most collagen is found and how to best utilise it and transform it into gelatine, whether for a jelly or for the unmatchable texture that beef cheeks possess when collagen becomes gelatine during a long, slow braise. A butcher will never be as good as they can be if they have no interest in the before or the after. It's as simple as that.

As a trade and a craft, butchery started out of the necessity to feed and nourish. An animal was born, fed, raised and harvested with respect and an inherent reverence for the life that would inevitably be taken. A carcass was utilised in its entirety, with nothing left on the waste pile. Let me say that again: *there was no waste*. Whether some of the inedible organs and less palatable pieces were used for fertilising and nourishing the land or whether some were eaten as is, or cooked or cured for the leaner times, it was all used. There was no real priority assigned to any one muscle, organ or bone as the entire carcass had a use. Clothing, fuel, food, tools, nourishment of body or land – it all had a purpose. We're a spoiled race now where we have choices everywhere, and I'm not suggesting that is a bad thing. I do, however, believe our vast choices have steered our ethical attention and mindset away from the essence of taking a life. Use it all, use it well and use it sparingly. That has to be the goal.

Aside from being ethically the right thing to do, whole-carcass utilisation gives the butcher access to every part of the body to be able to fabricate, showcase and talk about the many cuts they have in their display. Topside tender will rarely be seen in a shop using anything but carton meat, and petite tender (*teres major*) and velvet steak are an impossibility unless the butcher is buying cartons of heel muscle and matambre (the twitch muscle that moves rapidly under the hide of cattle to discourage flies, birds and other annoying animals from landing on their backs). Not to mention all that suet (visceral fat from around the kidneys) that can be transformed into the most magic liquid gold or Christmas pudding. The list goes on and on.

Aside from the greater flexibility and options the butcher of whole carcasses has, there is also the ability to make the traditionally expensive cuts cheaper. All those sweet cuts – sirloin, scotch, tenderloin – can be made a little more accessible as the butcher has the opportunity to showcase other cuts normally destined for the grinder in the display. The more carcass balance the butcher is able to achieve, the more they can spread the burden of cost across the carcass.

There is only one other important ingredient in this scenario, and it's a deal-breaker and a big part of why whole-carcass butchery began its fall from grace: the butcher needs buy-in from their customers. We need our customers to have the same fascination we have. They need to trust us and our recommendations. They need to know that we are the experts. And then we truly need to be the experts lest we demolish all the trust our customers have in us.

I'm not quite sure where the ridiculous notion that tender equals quality came from, but it has done nothing but damage to the cause of whole-animal butchery. No one can, hand on heart, say an eye fillet is better than a beef cheek if both are cooked appropriately and to their full potential. A Ferrari is no good on a dirt track just as a tenderloin is no good in a daube of beef or a beef shin on a grill. It's not rocket science. This view that tenderness equates to quality needs to be erased from people's minds and fast. Quality should in fact be dictated by the building blocks that made the muscle that sits on our chopping block and later on our plate. Genetics, care, feeding length and quality of life, and humane, calm and swift slaughter followed by immaculate post-death processing are what will ultimately write the rest of the story and dictate quality. Take short cuts on these and you've compromised the quality, and no butcher or chef can right these wrongs no matter how hard they try.

Another vital and welcome shift in opinion and practice over the past few years has dictated that we all – butcher and chefs – give more attention and focus where it belongs, to the producers. We are nothing without them and their product, and I believe we are merely (and I don't say that with any disrespect to other butchers) a conduit from them to the consumer. I'm not for a second suggesting butchers aren't important and our skills and knowledge aren't needed. Without butchers, the whole system will fall over and our suppliers and customers will be left with nothing. Supply-chain anarchy! I just feel that the true light should be shone on the people who breed, grow, feed and process our food. It is like a building without the floor. Without the foundations we have nothing, or at the very least we have a substandard and inferior product that not even the best chef or butcher can elevate to the high standards we expect and deserve. It's pretty simple in my eyes. Let the produce do the talking and we can help shout that story even louder.

THE PROBLEM WITH WET FISH

This is a controversial topic that inevitably brings about polarised viewpoints.

I think we need to start at the very beginning and explain that, firstly, a poorly killed fish or one that is neglected immediately post-capture can detrimentally affect the quality of the texture, shelf life, internal and external aesthetic quality and – above all – flavour.

Washing your fish in fresh water is a definitive *no* in the business of processing fish, and I'll explain why.

Water molecules move across semipermeable cell membranes through a process known as osmosis. They usually do this in order to achieve an equilibrium where the liquid on both sides of the membrane is exactly the same.

So when a saltwater fish comes into contact with fresh water, that fresh water will immediately start to penetrate the fish's cells in order to try to make them less salty. When this happens the cells can become so full of liquid that they rupture, causing the fish flesh to turn an opaque white colour. This not only degrades the texture of the fish but also its shelf life, as it creates an exposed and moist environment for bacteria to accumulate and thrive.

As this fresh water usually comes from a tap and is not at 0°C (30°F) either, this also causes a break in the cold-chain control of the fish, compromising its shelf life even further. And when you freeze a fillet in this condition, the water inside the cells expands as it turns to ice and ruptures many more cells, so when you defrost it you will have a compromised, mushy fillet.

Beyond the damage that washing a fish does to the fish itself, the impact of this process is far more detrimental to the aroma and aesthetic of the venue where the fish is being processed. Very rarely would you walk into a fish shop or market that didn't have a strong 'fishy' aroma; this will have been caused by the trimethylamine oxide in the fish breaking down and converting into ammonia, in turn creating the odour that we refer to as 'fishy', a breakdown promoted by excess washing, disruption of cold-chain control and rupturing of the fish's cell walls.

Remove the washing and wet-processing and not only will you have a fish that lasts longer, tastes better and is easier to cook with, but a market, fish shop or supermarket that has no smell at all.

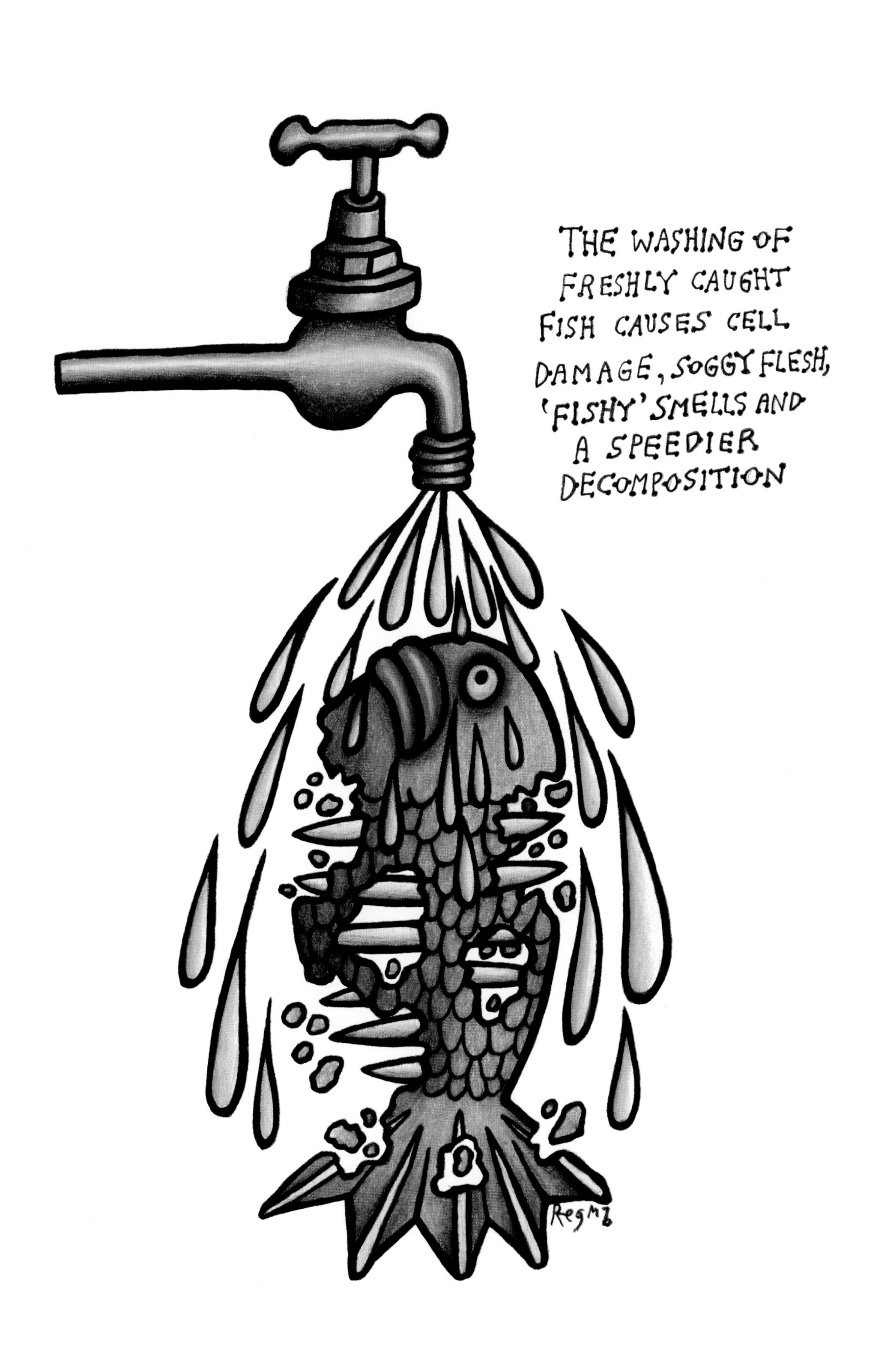
THE WASHING OF
FRESHLY CAUGHT
FISH CAUSES CELL
DAMAGE, SOGGY FLESH,
'FISHY' SMELLS AND
A SPEEDIER
DECOMPOSITION
RegMz

WHAT IS RIGOR MORTIS?

In fish, energy is stored in the form of adenosine triphosphate (ATP), which is produced using oxygen during the process of aerobic respiration. When required, ATP is then able to bind to muscles before breaking down into adenosine diphosphate (ADP) and phosphate, in the process releasing energy that is then used to power muscle contractions.

The remaining ADP molecule that is still bound to muscles is then replaced with a new ATP molecule, which causes the muscle to relax. The process then repeats itself as muscle contraction continues.

When fish die, obviously respiration no longer continues, which means that there is no longer oxygen available to produce new ATP. This means that there is no longer a constantly supply of ATP available to replace ADP, which results in the muscles of the fish being unable to relax.

While respiration would have stopped completely, electrical nerve messages from the brain and spinal cord signalling for muscle contractions do not cease. As the muscles continue to contract, an ever-increasing amount of ADP is not replaced, and from head to tail the fish starts to stiffen up in a process we know as rigor mortis. The process of rigor mortis takes time to set in, typically 16 to 24 hours in a fish, depending on its size.*

In a fish post-mortem, muscle contractions continue to occur without the supply of ATP from aerobic respiration. The dead fish is still able to produce ATP through the same process of anaerobic respiration in order to meet the ATP demand. This then results in the same build-up of lactic acid, causing a potential sourness in taste and a compromised texture due to the acid cooking the flesh from the inside out.

DRY-AGED FISH

The first time the penny dropped for me about the possibilities of fish-ageing was when I started working at Fish Face in Darlinghurst some 17 years ago. It was there that I remember being scolded one night for having forgotten to cover my portions of fish in the service fridge. Because they had sat under the fan all night, the skin had dried out. Despite being told that this would result in an inferior product because the fish had not been stored correctly, I cooked them and noticed that they were so much simpler to work with. The skin was far crunchier when pan-fried, the wetness of the exterior flesh had been partially dried and everything just seemed, well, easier.

It felt like I was cheating.

Years later I applied the same thinking to a whole fish and, rather than letting it lay on a tray, I hooked it from the tail, hung it and left it to age in a coolroom.[1] I haven't looked back since – when applied correctly, ageing a fish like this adds a profound amount of flavour, enhances the texture and composition of the flesh and results in a superior outcome when pan-frying or grilling.

Flavours and textures aside, the dry-ageing or correct preparation and storage of fish like this was also intended to arm my Saint Peter team and me with the ability to maintain an excellent day one-quality fish for an extended period in case we didn't sell what we purchased in a reasonable period of time. The result of this intentional storage method was that the fish performed significantly better. Customers started commenting on how delicious the fish was or, on some occasions, making comments like 'Wow, this fish is so beautiful and fresh' about a three-week-old tuna.

Since introducing this technique formally in my first book, *The Whole Fish Cookbook*, I want to touch on it again as I feel it deserves some points of clarity. When dry-ageing fish, all the following need to be considered:

1. It is critical that the fish you wish to work with has been caught, killed and transported correctly (see page 35), otherwise you are already on the back foot and no amount of ageing will result in an optimal outcome.

* The Japanese Food Lab (May 2019) 'What is Ikejime', The Japanese Food Lab, available at www.thejapanesefoodlab.com/ikejime, accessed 2 August 2022.

1. Hanging a larger fish from a hook like this and storing smaller fish on perforated trays/racks were both solutions to the issues around wet fish. Why did I hang the fish? The problem that I was trying to resolve was that when a fish lays on a tray for a day or even a number of hours, it will have moisture underneath it and the top side will start to dry if uncovered. To combat this, the fish can be turned each day to move the wet side up to dry and dry side down for moisture and so on. However, when moisture created by condensation accumulates, it can result in bacterial proliferation. So rather than purchasing fewer fish more frequently to avoid potential spoilage, the idea was to keep the fish completely dry throughout the whole process post-mortem.

2. A fish that's washed during processing (see page 28) will likewise never reach its full potential (not only in taste but also in flavour, aroma and texture).

3. Dry-ageing a fish takes place without the use of any salt or preservatives.

4. Dry-ageing is the process of intentional and controlled moisture loss to promote more of the fish's natural fats. Furthermore, maturing a fish on the bone will allow the development of glutamates within the fish, giving you a more savoury flavour profile and also a flavour that is identifiable from species to species.

5. Dry-ageing is a necessity for the fish industry to mitigate the gross amounts of fish waste produced every year.

It must also be said that a static (or almost static) coolroom is critical to achieve an optimal outcome.[2] The use of a fan-cooled coolroom will rapidly speed up the moisture loss of the fish without it developing any depth of flavour. There are now a multitude of commercial refrigeration options for storing fish and meat protein, and often you will find salt bricks in the base of refrigeration systems designed for ageing meat. Naturally, the salt bricks will draw moisture from the fish, so I understand their purpose, however from the numerous tests we have conducted, the results are as severe as placing the fish in a conventional fan-based coolroom. The simplest solution to this is to just take the salt bricks out.

It is the role of the fish butcher or chef to identify at what point the right amount of moisture has been removed from the fish. Some (most) would argue, 'How ridiculous, no one wants to eat a dry fish.' I couldn't agree more with that, however we are not talking about weeks and months of 'drying' to result in an emaciated fish jerky. This is controlled moisture loss with the intention to stop when the fish has reached the desired dryness. I know personally that I don't want to age a whiting past day four as it's just not the same memory-burning experience as it was when it was freshly landed. It is also important to note that certain undesirable qualities in a fish will become prominent even if the dry-ageing method is followed in the handling. A fish with a seaweed-rich diet that is matured too long, for example, can not only end up dry as it lacks considerable intramuscular fat but also deliver a flavour profile with an iodine metallic taste.

Everyone's optimum humidity for dry-ageing will differ based on their own personal theory and the quality of the fish, however we have found anywhere between 75 and 80 per cent to be our sweet spot. Again, this will be subject to a multitude of variables: size, species, amount of fat, method of catch, time of year, quality of refrigeration and handling – the list goes on. Depending on the fish species in question, the dry-ageing method we use in our larger rooms results in the loss of somewhere between 2 and 3 per cent moisture per day. Financially, this is a substantial cost, just like the maturation of beef or lamb, but it is worth it for the improvements in flavour, texture and superior pan-fry or grill results.

Lastly, it's worth considering the following: if a fish is delicious, does this mean it's fresh? And why should this matter? The argument I hear a lot is 'why not just eat the fish fresh?', and the answer is that all fish have certain moments where they will taste considerably better. This may mean, for example, that the window to consume the best sardines or anchovies in the world lies within the first 5 or 6 hours of them coming out of the water. Or that a turbot is at its most memorable in the first 48 hours out of the water but then peaks again with new-found flavour and texture some 10 days later with a reduced amount of moisture and more fat prominent. The idea is to manage and control the variables of fish catching, processing and storing to maximise the full potential of its shelf life.

A NOTE ON SCOMBROID (HISTAMINE) FOOD POISONING

Scombroid, or histamine, food poisoning can occur when an individual consumes certain types of fish containing high levels of histamine – a result of bacterial enzyme activity that has occurred in the fish post-death. The fish types affected are those from the scombroid family, including tuna, mackerel, bonito and kingfish (yellowtail), but also non-scombroid family fish including mahi-mahi, marlin and swordfish. The histamine can't be destroyed by cooking, smoking or freezing, with the only preventative measures that can remove or minimise the risk of this enzyme activity occurring relying on the fish being chilled soon after being caught and then being kept refrigerated until it is cooked, preserved or consumed. When handling these fish then, particular care in refrigeration is therefore necessary.**

---

2. Recently, Fish Butchery has moved into a far bigger commercial coolroom to store its fish, and while the room is predominantly a static space, the assistance of a fan functioning at 10 per cent capacity allows the room to maintain its cold temperature when being opened and closed. Without this, the room temperature would fluctuate too aggressively, and it would take far too long for it to cool down after opening. This was a system we needed to create if we were going to upscale our operation into something more commercially viable. We have a number of coolrooms across our businesses – the smaller two-door systems allow us to work with a very limited number of products and perform well because they are not trafficked as frequently as our bigger spaces. The rooms all have rails for hanging the larger fish and shelving for perforated gastronomes or racking for smaller fish.

** Queensland Government (June 2021) 'Scombroid (histamine) food poisoning, Queensland Government, available at https://www.qld.gov.au/health/staying-healthy/food-pantry/food-safety-for-consumers/potentially-hazardous-foods-processes/seafood/scrombroid-histamine-food-poisioning, accessed 2 August 2022

A FISH'S JOURNEY FROM SEA TO STORE

While it might seem somewhat contrary to finish this section with the very beginning of a fish's journey to our plates, nothing is more important in determining product quality within the fish industry than an understanding of the 'best practice' processes involved in landing the fish in the first place and getting them to their final destination. For all the preceding conversation around fish butchery, handling, dry-ageing and preparation, unless this first step in the chain is adhered to, an inferior result will be all but guaranteed.

For line-caught fish, the process for a fisher is to bait up, throw a line in and wait until the fish jumps on to the hook. When the fish can't circulate the amount of oxygen it needs to resist, it gives up the fight. While trying to escape, the fish has developed a significant amount of lactic acid in its muscles. At this point it is gold standard for a fisher to:

1. **Spike the fish in the head to inflict an immediate death.** Allowing a fish to thrash around and die slowly in a bucket of water or open container is unacceptable – it causes a tremendous amount of unnecessary stress to the fish, reduces the eating quality and shortens the shelf life of the flesh. Spiking of the brain (originally a Japanese technique called *ikijime* or *ikejime*, pronounced 'iki-jimi') will kill fish immediately; it should be done quickly – preferably within a minute of it being caught.

2. **Clip the gills to bleed the fish.** If the fish is not bled correctly then the blood that resides within the fish will return to the flesh. This causes a number of issues: it dramatically shortens the shelf life of the fish, it visually alters the colour of the flesh and the flavour of the fish is compromised where it starts to carry a heavy metallic taint.

3. **Place the fish in an ice slurry to further remove lactic acid as well as reduce the temperature of the fish.** It's then critical that the fish is transported correctly from catch to market, keeping it between 0 and 1°C (30°F). This can be done through the use of ice packs surrounding the bottom, sides and top of the fish. Alternatively, if shaved ice needs to be used then keep this separate from the actual fish by placing an impermeable sheet between fish and ice.

At this point, the individuals who are purchasing the fish that have been caught and killed need to have established what they are doing with the whole fish. What's the game plan? There is no point purchasing a fish only for its fillets as this is exactly where the problem lies. Upon arriving at the market or store, the fish needs to be removed from the packaging into a chilled environment to ensure it stays as cold as possible. Then the purchaser needs to prepare the fish as follows:

4. **Scale the fish, (see page 49) using the method best suited to the fish size.** This might mean using a spoon for very small fish, a small head scaler for a plate-sized fish or a variety of sized ring scalers for larger fish. Larger fish can also be scaled with a sharp knife. While most of the time scales will make up a very small percentage of waste, they too can be cooked a number of different ways to achieve a unique textural outcome. Once the scales are removed, the fish must be wiped with paper towel or a cloth to ensure that none remain. The temptation here is to wash the fish and the surrounding surfaces down thoroughly to speed the process up, but resist as this is where the issues begin (see pages 28).

5. **Remove the organs from the fish.** 'Gutting' conjures images of waste, blood and mess to clean up, but a correctly gutted fish should be a simple, quick and clean process. The art lies in the selection of sharp tools, including scissors and a straight-edge (non-flex) knife. A fish can be gutted in a number of different ways (as demonstrated on page 49), but be mindful that if your knife is inserted too deeply into the cavity of a fish then you will puncture the organs within, which makes it more challenging to use them in other preparations and creates a lot of mess to clean up afterwards. Once the organs and gills are removed, discard the gills and gall bladder as these (at least to me) have very little culinary application. At this stage, wipe the cavity of the fish completely clear of any blood or residue using a cloth or paper towel, but DON'T WASH THE FISH.

   Just as the exterior quality points of a fish are considered, the interior elements must also be – if the offal inside the fish is compromised in any way then unfortunately this will make up a part of the waste or need to be considered in other ways. The challenge now is sorting through the usable organs of the fish and actioning immediate conversions to them; this might mean salting the heart, spleen, intestines, stomach and kidneys in readiness for applications the following week or removing the liver and trimming it in readiness to make pâté or simply pan-frying for salads or to serve on toast. The other part of the fish that is sometimes regarded as offal is the head, a somewhat fragile piece of the puzzle that contains soft organs like the eyes and brain and requires some intentional thought. The head can remain on the fish to make it a part of the preparation (see, for example, page 92, where the head is split so the fish can be grilled) or it can be removed then steamed or simply cooked with the intention to pick the meat from it for another purpose. Another very niche concept (yet fascinating all the same) is to use the vitreous humour in the eyes in ice cream (see page 253).

6. **Consider your timeframe.** If, at the market or store, it is known that the fish at hand will not be purchased or used in the next few days then the fillets will be left on the bone. Removing the fillets from the bone exposes the flesh to light and oxygen, which will affect its taste and texture. Leaving it on the bone starts to develop its natural glutamates, which make the fish taste more savoury and remarkably more delicious. This then feeds into the conversation around dry-ageing and intentional moisture loss in a fish to create desirable flavours and textures.[3]

7. **See pages 79–149 for cuts.**

3. This is not a conversation about putting a fish into a coolroom for a month and it somehow becoming significantly better; this is the empowering knowledge that if you handle a fish correctly from step 1 to step 7 then you will be working with a far superior product that you have yielded more from and have more time with. Some fish are so much better on days one, two and three, and I would suggest using your instincts to make this judgement. However, throwing a perfectly good fish in the bin after three days or selling a badly handled fish to an unassuming consumer who will have a poor experience is just sending the whole industry backwards, along with adding to the amount of food wasted every year.

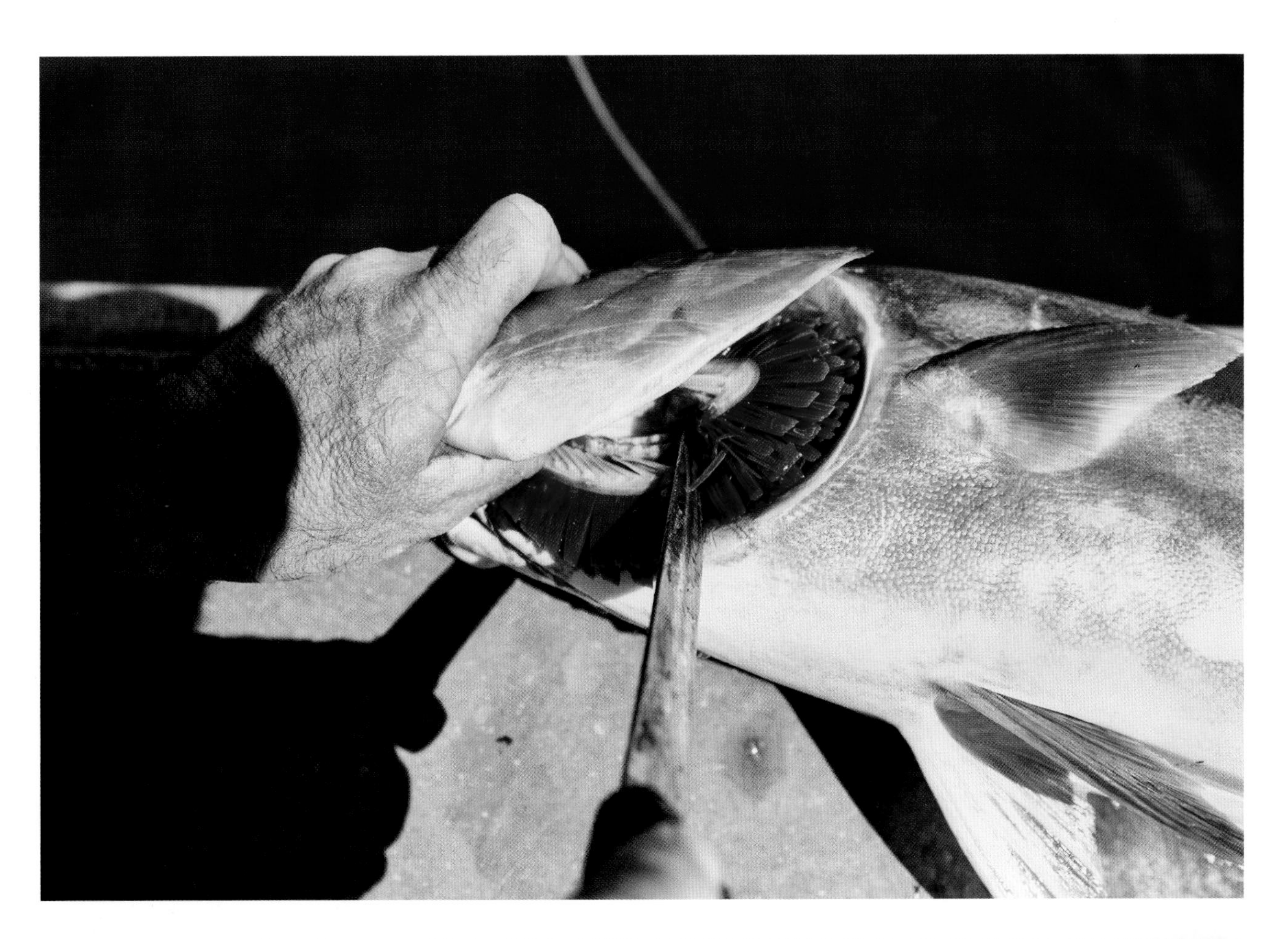

BRIXTON

THE FISHER

A NOTE FROM LUKE BUCHHOLZ, COMMERCIAL FISHER

Commercial fishing for us was not something passed down over generations, so when my brother and I decided to start we didn't have any bad habits. We came with a fresh set of eyes and, because we weren't the best fishers, we needed to get maximum return from what we caught.

We didn't realise it at the time but our practices of bleeding and *ikijime* spike killing were things I had learned 15 years earlier when I was a keen spearfisherman. We would kill our fish as soon as possible so that the sharks wouldn't come in, and we'd bleed them straight away by gill clipping, because that's what the guy I had learned under was doing.

We also noticed that the fish tasted significantly better when they were killed with a kill shot instantly. The fish didn't have time to stress or build up lactic acid from fighting on the line.

I found it strange that this wasn't normal practice with commercial fishers – it was all about the 'kilos', so to speak. We also noticed that fish in the shopfronts didn't look like our fish. They were knocked around, with scales missing or white sunken eyes. We had started out as lobster fishers – trapping fish in between pulling lobster traps – and sold all our fish through the local co-op, which we found very frustrating as the majority of our fish were sent to markets. If multiple other fishers weighed in the same species of fish we would *all* get paid the same amount, irrespective of the difference in quality of our fish at the other end.

So we decided to go out on our own and do things our way.

We started shopping fish around the country using road transport, which was another issue for us. There was nothing worse than our fish getting lost or put on the wrong truck or being told that we didn't have enough ice on our fish even though when it left us the boxes were full, which meant our fish had been left out somewhere and our ice had melted. To combat this, we took the transport side into our own hands as well. We do our best to get our product from the water to the end consumer as quickly as possible (usually between 24 and 48 hours, but sometimes as fast as 3 hours for our local chefs).

Instagram was our social media platform and that really took things to the next level with new customers. Having direct contact with chefs took the middlemen out of the equation which essentially doubled – if not tripled – our returns. The fluctuations of fish prices at the markets were always really hard to predict and, irrespective of the quality of catch, liable to change overnight. We noticed that as soon as a fish comes into season, like bonito, for example, the price on the market floor would drop from 25 dollars a kilo to 5 dollars a kilo over the space of a week and stay there for the remainder of the season.

By having direct contact with our chefs we were able to set a year-round flat rate, which worked out better for both us and our chefs as they knew what to expect before ordering. The set-up and running costs of doing everything in-house do add up and can be broken down like this (prices listed in Australian dollars):

- ice machines (we have two) – $10,000 each
- coolroom – $12,000
- delivery truck– $35,000
- processing/packing facility – $40,000
- other equipment (scales, fish boxes etc.) – $2000
- running costs (for the ice machines and coolrooms) – $750 per quarter

- maintenance and servicing on the delivery truck – $200 per month
- fish receiver's licence – $4000 per year
- food-safe licence – $2000 per year
- insurance on equipment and delivery vehicle – $2000 per year.

Like in any business, there are always upfront and one-off costs, but that is still money that needs to be accounted for and recovered at some stage.

There are other factors that can affect your catch rate. The extra time it takes to carry out the kill methods mentioned above as opposed to just throwing fish directly into an ice slurry can reduce your catch rate by up to 50 per cent. Take bonito fishing, for instance. If two boats were fishing side by side using the same fishing techniques, lures and bait, then our catch using these kill methods would be 25 to 50 per cent less at the end of the day and we would have to stay back an extra hour or two to fill the same number of orders as the other boat.

There is also the extra workload that comes with what we do. About 90 per cent of commercial fishers drop their catch off at a local co-op or wholesaler and get paid the following week. After catching our fish, we have to start the process of contacting our chefs and letting them know what we have. As not every chef orders every time we have fish (they may have a function on, they may have already ordered their fish, or they simply don't want the fish that we have), you need an extensive range of clients to be able to move volumes of fish, and building our customer base has taken time and patience. Once all the orders are in we head out to our packing shed to prepare the orders and load up our delivery truck. That's followed by invoices and a delivery plan for our driver (who can sometimes be my brother, my wife or me) to use the following day. A run can take up to 15 hours to complete, then the delivery truck is cleaned and readied for the next run. It doesn't stop there, though, because we then have to chase up accounts and reconcile outstanding invoices that can affect a small business's cash flow.

There are other issues that can slow down orders too, including the day the fish is caught and the day it is delivered (most chefs want their fish on Tuesday or Wednesday for prep), the amount of fish caught (for us to deliver from the mid-north coast of NSW to, say, the Gold Coast or Sydney, we need at least 100 kilograms of orders to be financially viable). We try not to fish over the weekend as we have found it's extremely difficult to move fish at that time, and I don't like sending fish that has been sitting around for three days to the markets.

We plan to do the majority of our fishing from Monday to Thursday each week, but we still have to factor in weather, tides, fishing conditions and outside influences like recreational fishermen or extreme events like cyclones or floods (our river has flooded three times this year alone, which basically stops all fishing for at least two weeks). And sometimes the fish just don't want to play ball, which adds an extra stress.

To summarise the whole picture, the extra workload involved in doing things our way would be at least 100 per cent more than the industry standard of dropping your fish off at the local co-op and getting paid the next week, but financially we are around 150 to 200 per cent better off than those businesses. It's not all about numbers, though – we are extremely proud of what we do and how we do it, and most of all we love the enjoyment that chefs and consumers get from our fish. I want this business to be here for my kids and their kids, so I feel we have a duty of care to provide for them by doing the right thing, looking after every fish that we pull from the ocean and passing on our knowledge so they can keep improving on our fishing methods and reduce the impact we have on our oceans and wildlife.

CUT

Whether it's the head, tail or offal contained within, a fish presents so much more culinary possibility than simply that central fillet cut. There is incredible opportunity within a single fish to realise not only a more desirable and delicious outcome but one that is intentionally seeking the greatest possible yield and minimising what goes in the bin

This section outlines the constituent parts of a fish, the tools needed to process one efficiently, the steps involved in preparing a fish for butchery and a variety of the cuts into which they can be transformed, with the aim of minimising waste and treating the whole animal with respect. Simply put, our approach to processing a fish should be the same as how a butcher would break down a pig. Why do we think it's okay to throw half a fish in the bin?

Intelligence and intentionality will be what saves our global fish stocks.

TOOLS OF THE TRADE

KNIVES

The choice of knives is so much subject to the individual and their preferences. I, for example, only like to work with knives that have no flexibility in them at all, but this is a personal choice. Below are some of my favourite knives and the reason I use them.

Chef's Knife (230 MM BLADE) Excellent for all-purpose cutting of vegetables, portioning fish and cutting the scales off a fish.

Long Sabatier Knife (230 MM BLADE) This knife is my perfect filleting tool for larger fish species as it allows me to efficiently make my fillet in three or four cuts due mainly to its length and durable hard steel. I also use this knife to portion fillets.

Short Knife (130 MM BLADE) Having this knife to hand is important for fish like herring, mackerel, sardines and even smaller plate-size fish. Smaller knives offer a lot of control and speed during filleting but are also a great tool for gutting and knife-scaling a fish.

Cleaver (170 MM BLADE) Just as a meat butcher's knife belt includes a cleaver, I feel a fish butcher needs one within arm's reach for assistance when splitting heads in half and getting through large vertebrae. Sharp, not too large and with a short handle.

SCISSORS

(SILKY OR CHIKAMASA, 100 MM BLADE)
If there is one item in our artillery that I am still on the hunt for it would be a great pair of scissors. Although we have over 50 pairs at Fish Butchery, we still haven't found the ones we love. Many have come close, but when working in large volumes you need reliable, sharp, comfortable and lightweight scissors that cut through large and small bones with ease. Scissors greatly speed up our efficiency when filleting and doing more intricate butterflying and detailed cuts of fish. Invest in a good pair!

PLIERS AND TWEEZERS

Every bit as important as scissors, pliers and tweezers play a critical role in fish processing in order to remove large and small bones with ease. Using tweezers for a job that needs pliers can often double the amount of time needed to complete a task.

MINCER

A hugely important piece of equipment for a commercial fish business to have to hand, a mincer will allow you to convert the scraps and less desirable parts of the fish into beautiful fish mince for sausages, patties and more. Half the repertoire of a butcher of land-based animals wouldn't exist without a mincer, so why not fish?

MEAT SLICER

A good meat slicer is an invaluable piece of kit that pays for itself very quickly. We use it religiously for slicing our mortadella, bacon, hams and cured fat.

HOOKS

S-shaped butcher's hooks are a critical piece of equipment at Fish Butchery as they grant us the ability to hang our fish. This prevents the fish protein from coming into contact with another fish and allows the moisture that develops on the surface to evaporate, preventing the development of strong aromas, keeping the flesh from spoiling and giving the skin the opportunity to begin to condition itself with the assistance of the static environment the fish is housed in. Hooks also come in handy when it comes time to roast a whole fish or bone-in cut (when roasting fish on a tray, the bottom fillet always suffers the fate of being overcooked while the top is perfect – hanging a fish securely while roasting allows the skin to crisp all the way around to achieve a texturally consistent result).

FISH WEIGHT

To me, you can't grill or pan-fry a fish well without the assistance of a fish weight. Whether it's our Saint Peter fish weight, a fish press, burger press, a saucepan filled with water, another frying pan or even a brick, the weight is essential for pressing the skin to the base of the pan and creating an even surface area, giving you the best chance of achieving a crisp golden or evenly blistered skin when pan-frying or grilling.
As heat rises around the portion of fish that you are cooking, a specifically stainless steel fish weight will also capture the heat and begin to conduct that warmth across the flesh side of the fish, meaning the cooking time is reduced and the silky elegant texture of the fish remains.[4]

SCALES

Having a reliable, accurate set of kitchen scales is critical when portioning fish. Even a 10 to 20 g (¼ to ¾ oz) variance between portions can not only frustrate paying customers if they feel they are short-changed, but can also swing the other way and start to affect the financial sustainability of your business if portions are overcut. On top of that, it's essential to have a large set of scales that can determine the weight of the whole fish being received so you can work out yields and costs before the knife even goes in.

DIGITAL PROBE THERMOMETER

A critical piece of equipment that you must have in the kitchen when working with both raw and cooked fish. This is to ensure not only safe practices and manage the cold chain controls of a fish, but also to achieve desirable textures and stable temperatures of fish protein during cooking.

TRE SPADE SAUSAGE FILLER

We use this hand-cranked piece of kit to fill all the sausages we produce at Fish Butchery. To us this is an essential piece of equipment that really brings efficiency to our sausage and wider charcuterie program.

OX RUNNERS (BEEF CASINGS)

While it would be wonderful if we could use fish intestines to house our sausages and salamis (and trust me, I've tried) beef casings are significantly thicker and far more shelf stable, making them an excellent choice.

PACOJET

This piece of equipment helps us to produce refined, brightly coloured herb oils and ultra-smooth rendered fats, as well as to bring our fish eye ice cream (see page 253) to reality.

ROBOT COUPE

We use a Robot Coupe to achieve the specific textures we need within our mortadella and frankfurt recipes (see pages 158 and 185).

COMBINATION STEAM AND CONVECTION OVEN

A combination steam and convection oven (we use a Rational Combi) gives us complete control over temperature and humidity, allowing us to steam and roast at low temperatures. Whichever oven you use, follow the suggested times, oven temperature and internal temperature to achieve the same outcome.

SMOKERS

We use two types of smokers at Fish Butchery, one is the Bradley brand 6 rack digital smoker that is capable of cold and hot smoking. When opening our Waterloo site we needed to get a larger system so we purchased a 4 litre Smokai brand smoke generator which we retrofitted to a standup commercial refrigerator. By doing this we gave ourselves the ability to cold smoke under fully refrigerated conditions.

CRYOVAC MACHINE

Essential for the storage of a number of our charcuterie products. Once sealed in cryovac bags, there is less risk of contamination, along with a reduced risk of further moisture loss if the desired texture has already been reached.

VITA-PREP JUG BLENDERS

We use a number of different blenders, food processors and grinders throughout the butchery, however, when we are looking to achieve a very fine-textured outcome we turn to a jug blender. A high-powered blender, it produces the smooth results we need for preparations such as our liver pâté (see page 193).

4. One of the biggest challenges cooks face both domestically and professionally is the execution of a pan-fried or grilled piece of fish. While a wet fish will always be a challenge to work with on direct high-heat methods of cookery like grilling and pan-frying, finishing the cooking process is tricky regardless of whether your fish is wet or dry-aged. If a weight isn't used during cooking then although the skin might be crisp or well-coloured, how do you finish the cooking? Many believe that in the oven, skin side down is the best way, which I agree is an excellent solution, however what happens more often than not is butter will be added to the pan, the portion turned flesh-side down and the browning butter spooned over the skin to reinforce the colouring and to add flavour. Each to their own, but for a memorable experience, keep the flesh of the fish protected from the direct heat by the skin and notice the extraordinary texture of the gently set flesh.

K Sabatier

SCALING, KNIFE-SCALING AND GUTTING

Cutting the scales off a fish allows us to condition the skin more intentionally as it no longer harbours any unnecessary moisture. A sharp knife passing over the surface of a fish does little to no damage to the condition of the flesh. The knife-scaling technique depicted visually here is a traditional method from Japan that aims to minimise bruising and textural differences across the flesh of a fish that is intended to be used for sushi and sashimi.

Too often when a fish is gutted, the whole length of the blade is inserted and the contents of the fish destroyed for a lack of intentionality or consideration that it has any potential that exceeds becoming fertiliser. Take care not to puncture any of the offal when gutting – the recipes that follow will give an insight into what can be achieved.

SCALING

Flat fish such as flounder and small fish such as whiting, garfish, herring, sardines and snapper can be scaled with a small knife, small-headed fish scaler or spoon.

1. Gently run the scaler from the tail to the head, working methodically around the body of the fish and applying only enough pressure to remove the scales.

2. Continue until you are confident all the scales are removed, then wipe the fish with paper towel.

SUKIBIKI (KNIFE-SCALING)

For large fish (plate-size or bigger), scales can be cut off if your knife skills are up for this.

1. Firstly, using a pair of kitchen scissors, cut all the fins off the fish that will impede your ability to cut the scales off easily.

2. With a sharp knife that can be either short or long, depending on your comfort, start at the tail end of the fish and hold your knife almost directly parallel to the bench while pressing against the scales/skin of the fish. Angle your knife very slightly to allow the blade to slip between the scales and the skin, then, using a back-and-forth motion, begin to cut away the scales in long strips.

The aim is to remove the scales and the membrane that carries them, along with natural moisture present, while leaving the skin of the fish intact. The first few times you do this you may puncture the skin, exposing the fish's flesh. Don't panic, simply correct your knife and continue.

GUTTING

This is the traditional method of gutting, which to me is the cleanest and simplest. However, in some of the cuts that follow there will be other methods of gutting that start in a different position and enable other unique cuts to be achieved.

1. To gut the fish, make an incision at the fish's anal vent. Using only the very tip of a sharp knife, cut up through to the gills under the bottom jaw of the fish.

2. Once the cavity is opened, use scissors to cut the membrane that sits in front and behind the gills and collar.

3. The gills can now be pulled down towards the tail, and the internal organs of the fish can be removed in one piece with little mess.

4. Wipe the cavity and skin of the fish very clean with paper towel. Reserve the offal.

OFFAL UTILISATION

Before you start thinking about offal utilisation, understanding where your fish is from, who caught it and how long it's been out of the water is critical. This information will give you confidence in the knowledge that your fish was meticulously handled in unpolluted waters and that using the offal for cooking is a good idea.

Just as with the quality points of the external condition of a fish, we must look within for the same details. It is so important to ensure that only the very best and freshest offal is used from a fish. It is one thing to minimise our waste when processing a fish, but serve someone an inferior piece of fish offal as their first experience and you can only imagine how disappointed they will be. And remember, the importance of dry-handling the whole fish explained in the early stages of the book extends to this section as well – wet offal is about as useful as a wet fillet. By keeping the organs dry and free from water, the taste, aromas and textures will be far better.

While many modern diners hold the offal of land-based animals in high esteem, some may not be as willing to celebrate these texturally challenging cuts in a fish or even be aware that such an opportunity exists. The following pages showcase the various offal that exist within each fish along with some suggestions as to the quality points, extraction and possibilities.

HEAD AND COLLARS

In my opinion, leaving the head and collars (the cut along the fish clavicle, right behind the gills) on smaller fish and then proceeding to either butterfly or split them in half (see pages 92 and 84) are some of the nicest ways to serve a fish, as it offers the diner a multitude of tastes, textures and also delivers a sense of reverence towards the fish that's being eaten. However, not everyone wants to be looking into the eye of what they are eating, and the flesh that can be found within the head and collars can also be put to great use elsewhere.

Quality Points

- Firstly, ensure the head you are working with is of known origin. Like land-based animal offal, it is incredibly important that the product is extremely fresh and has been well looked after up until this point.
- The eyes are always a telltale sign of how fresh the fish is. Sunken, foggy, dry or even sticky eyes on a 'fresh' fish should be avoided entirely.
- The flesh on and around the collars should carry little to no aroma.
- The flesh on the head and collars of a fish should be vibrant, glassy and firm to the touch without signs of deterioration or discolouration.

Extraction

- Removing the head as a single unit can be simply done with a sharp heavy kitchen knife or cleaver, where the initial cut is made between the vertebrae on the spine. Finding this gap in the bones allows you to cut through with ease.
- The collars can be taken off separate to the head by making a diagonal cut from the top of the shoulder of the fillet down to the belly of the fish just behind the first fin (see step-by-step breakdown on page 82). Once the collars are removed, further work can be carried out to remove the smaller, more intricate bones to allow the diner complete enjoyment of boneless flesh with only one large structural bone to hold onto as they eat.

Application

- Some of the simplest ways to work with the head and collars, as I mentioned, are to leave them intact but present the fish either butterflied or split in half, ready to be grilled over coals, pan-fried or steamed, depending on the species. From a culinary point of view, the head is quite a forgiving component of the fish as it can tolerate intensities of heat without drying out as badly as the fillet, due in part to the amount of connective tissues, collagen and fats that are present. Because of this added fat and collagen, the head and collars, although perfect cooked simply with salt alone, can also be a great vehicle for sweet, sour and umami-rich marinades that add a significant depth of flavour. Cooking over coals is an excellent choice for these cuts. As the pores of the fish heat, they open and the fat begins to drip from the skin and bones onto the hot coals. The smoke from the coals travels back up towards the head and enters into the pores.
- For a more processed approach to utilising the head of a fish, remove the head from the frame of the fish and set aside. Once the heads of the species have been collected, simply steam these until the flesh that surrounds the cartilage and bones just begins to come away with the push of your finger. While the heads are warm but have rested for half the amount of total cooking time, begin to pick the meat. Like picking crabmeat, the priority is to keep the meat as large as possible because inevitably there will be scale, bone or cartilage missed and you will need to pick through again. Being too heavy-handed with this task will result in mushy, soft and small pieces of fish meat. The sky's the limit now with this unctuous head meat containing the cheeks and jowl of the fish. It is fantastic for the Fish Fingers on page 218, the Fishcakes on page 221, or even warmed through pasta, on toast or in a salad of crisp lettuce, raw onions, chives and Chardonnay dressing.
- The brain within the head of a fish is a very miniscule organ that I've never seriously pursued applications for on its own beyond once removing and tempura-frying it (which, to be fair, was pretty delicious). The brain content can be removed at the point of picking the meat off the head and is then a fine inclusion to a fish head terrine (see page 189), adding viscosity and richness.

A NOTE ON HEAD MEAT

A number of recipes throughout this book require the use of cooked and picked head meat taken from the head and collars of the fish. In these recipes we have started with a whole 3 kg (6 lb 10 oz) fish and on day one cut the head and collars off, weighing about 725 g (1 lb 9 oz). Place these on a perforated steamer tray or a steamer basket, depending on the amount you need and equipment you have, and steam in a Rational Combi oven preheated to 70°C (158°F). The time in the oven will depend on the species as well as the size of the heads and collars: we start at 12–15 minutes for smaller heads from 2–3 kg (4 lb 6 oz–6 lb 10 oz) fish and go up to around 30–40 minutes for large heads from 10–15 kg (22–33 lb) fish. Cook in batches, as you want to be picking the meat while still warm.

To check if the heads are cooked, test the cheek meat, which sits just above where the spine begins – the meat here should be opaque and flake away from the bone easily. Once this is the case, remove the heads from the steamer and allow to cool for a moment. Assemble two large bowls to pick into.

Once the heads are cool enough to handle, start extracting the meat. This can be a tedious task at first as you familiarise yourself with the structure of the head. I like to begin with the collars and get them all out of the way first. Gently peel away the skin then start to pull away any bones you can feel, discarding them into one of the two bowls. Try to hold the collar gently and lightly feel for where the bones are. The remaining meat should now be boneless and skinless, so place this into the second bowl, which will become your meat bowl. To pick the heads, begin again by peeling away any skin, particularly from the top of the forehead and the jowls, and place into your discard bowl. As the meat is exposed, use your finger to gently pop pieces away from the head and place into the meat bowl. Once the jowl and forehead have been picked, invert the head and begin extracting any meat from the inside. Continue until you have picked all the meat. Discard the contents of the other bowl.

As a final check, collect another bowl and, bit by bit, pick over the meat into the clean bowl, checking for any small bones, scales, tainted meat or skin.

LIVER

The first time I ever knowingly saw a fish liver was when I cut into a 2 kg (4 lb 6 oz) John Dory and plucked out a liver that was 210 g (7½ oz). That was over 10 per cent of the fish body weight! An average fillet of fish on a restaurant menu or a portion you buy to take home can weigh between 160 and 180 g (5½ and 6½ oz), so when the liver and other organs surpass this, the cost of discarded offal can be hugely damaging to fish businesses both financially and ethically.

Quality Points

- When inspecting the liver, there should be no visible signs of damage or discolouration. If the gall bladder of the fish has been punctured and stained the liver with a fluorescent yellow-green liquid, this can be simply trimmed off as it only affects the surface of the liver.
- The presence of worms and parasites can be the major cause of an inability to work with fish liver. Unfortunately, if a liver is contaminated in this way then there is no use for it at all.
- Livers that have sat in the cavity of a fish too long post-mortem can be subject to deterioration caused by the acidity of the stomach and intestinal contents. A liver like this can perhaps present well on the surface but when touched will almost deflate or turn pasty or creamy.
- A perfect liver will be firm to the touch, bright and uniformly coloured and have very little odour at all. Late winter and early spring in Australia is when we begin to see the best livers appearing.

Extraction

A liver is most easily removed from the cavity of a fish by gutting it conventionally in one piece with little mess (see page 49). Once removed, set the fish aside and position the organs on the cutting board as they were in the fish. Just beneath the gills will be the small fish heart, followed by the liver. Using sharp scissors or a small knife, cut the two main arteries off the top of the liver as well as the one underneath that connects it to the stomach of the fish. Once removed, inspect the liver and trim any significant arteries or veins that are visible.

Application

- The easiest way to cook a fish liver is to simply pan-fry it in a hot cast-iron skillet with a little ghee or neutral cooking oil and season liberally with salt and pepper. By leaving the fish liver pink inside you will taste the quality of the fats and experience its silky-smooth texture. This simply cooked liver can be a perfect addition to salads and pasta or – my favourite – enjoyed on toast with wilted parsley.
- Too high a heat when working with liver can often result in scorching the organ and creating a bitter aftertaste, while cooking a liver for too long will dull its flavour and render its texture a desiccated, powdery pulp.
- Liver terrines, pâtés and sausages are all incredibly delicious products that can be made with the right livers (see examples on pages 186 and 193). If you ever find yourself seeking inspiration for a fish liver application, look for a recipe using duck or chicken liver as they are comparable in size, texture and richness.

HEART, SPLEEN AND KIDNEYS

I have put these three organs into the one basket due to their richness and strength of taste. Each of them possesses a very oxygenated flavour profile that reminds you of eating meat offal – something to be harnessed rather than feared. These dark, blood-rich organs must be worked with astutely; their flavour can be sweet and their texture firm, however poor handling or keeping them for too long can result in a metallic taste, a sour flavour and a slimy texture.

Quality Points

- Look for unpunctured, brightly coloured organs that carry little to no aroma and a thin covering of fat, and are slightly dry to the touch.
- Blood will be the first thing to taint in these organs – the first visible signs of this will usually be browning and discolouration.

Extraction

The heart, spleen and kidneys can best be removed from the cavity of a fish by gutting it conventionally in one piece with little mess (see page 49). Once removed, set the fish aside and position the organs on the cutting board as they were in the fish. Just beneath the gills will be the small fish heart. This can be carefully snipped from the throat of the fish and set aside. Next is the spleen, which will either be found entangled in the fish fat that coats the surrounding organs or sitting in behind the intestines. It is very dark in colour and may only be as big as a thumbnail, depending on species. The kidneys are to be found where you see a dense mass of congealed blood nearly fixed to the spine of the fish midway down the bone. They can be removed with fish pliers.

Application

While there are a multitude of ways this offal can be used, I tend to prefer two methods that unlock a variety of different end results.

- The first is salting, which allows you the luxury of time and also the ability to strengthen the flavour profile of these organs (the other good thing about this method is that as you process more fish, these organs can be accumulated with confidence that they won't spoil). To salt the organs, simply bury them in regular fine cooking salt for a minimum of 7 days until hardened and cured. At this point you can dry them out completely in a low oven or dehydrator and then store them in an airtight container. This dried product can then be blended into a powder and used to make condiments, added to flours for baking, used as a seasoning for fried items or for providing a surprisingly umami-rich addition to egg dishes.
- The other application is to cut these three organs from the fish and cook them that day on a skewer over coals. This is an indirect method of heat application that will perfume the offal with smoke and cook them gently, preventing them from toughening up and becoming dry and powdery, a risk you would run if pan-frying them in oil. When cooking hearts and spleens specifically, make sure to leave them pink when cooking, use a good amount of salt and remember when seasoning that sweetness and acidity go exceedingly well with them.

ROE OR MILT

From caviar to bottarga and taramasalata, the roe sacs – or fully ripe egg sacs of a female fish – are the starting point for some of the most sought-after products in Western cookery. Call it good marketing or just plain delicious, this organ seems to have done okay in terms of being utilised thoughtfully. The problem here arises when the fish itself starts to be discarded in favour of its eggs.

Conversely, milt – or the sperm-filled reproductive gland of a male fish – can be seen as one of the most challenging ingredients to cook and consume. While the Japanese see *shirako* or fish milt as having extremely good anti-ageing properties and tuna milt is very popular in Sicilian pasta dishes, it hasn't quite reached the same level of enthusiastic uptake here in Australia. Yet.

Quality Points

- When processing roe, look for a sac that is uniformly and brightly coloured. In addition, the sac needs to be firm, unpunctured and without odour or excessive slime.
- For milt, quality points can be identified by its uniform bone-white appearance, little to no odour and even texture.
- Similarly to the liver, the roe and the milt can suffer the stain of a punctured gall bladder. Again, if this can be trimmed away then there should be no issue unless the intention is to cure the roe whole.

Extraction

The roe or milt can best be removed from the cavity of a fish by gutting the fish conventionally in one piece with little mess (see page 49). Once removed, set the fish aside and position the organs on the cutting board as they were in the fish. The roe or milt will sit together as a pair in behind the intestine and stomach. By snipping the connecting sinews, the roe or milt should come away quite simply from the remaining organs. Be careful not to puncture the membrane and gently remove in one piece.

Application

- For fish upwards of 2 kg (4 lb 6 oz), one of my favourite preparations of fish roe is to use a knife to cut the membrane open and scrape out the eggs within, then salt them to yield a result reminiscent of caviar (see page 227 for details). Once this product is ready it can be stored in this form in an airtight container for at least 1 week, while in a sterilised and sealed tin it will last for up to 3 weeks. I love adding this to raw fish dishes or even as an addition to an oyster.
- Alternatively, I like to cure and lightly smoke fish roe (see page 228) to create a product that is not technically a bottarga but still retains some chewiness and is excellent cut into thin slices and served with some good extra-virgin olive oil.
- For milt, I find that thinking of it in a similar vein to meat offal such as brains or sweetbreads overcomes any initial hesitancy of how to approach this organ, and the culinary opportunities become more apparent. Curing, smoking and drying milt (see page 233) results in a delicious product that is good sliced and pan-fried on toast or diced and added to smallgoods such as mortadella, sausages or terrines.

STOMACH AND INTESTINE

Stomachs and intestines might just be the most challenging components of a fish for people to get excited about eating, and I'll gladly put my hand up and say it took me some time to think of a desirable solution! When prepared correctly, the stomach is void of any real flavour at all, yet just like beef tendons or pigs' trotters, it can be used in dishes for its remarkable texture and ability to carry other flavours, while the tender intestine can be used in a variety of different ways. The quality points of all fish offal are incredibly important, but I can't stress enough that these particular organs must be exceptionally fresh and unspoiled in any way and are put to work inside of the first 24 hours.

Quality Points

- The stomach and intestine must be extremely fresh, clean-smelling, undamaged and evenly coloured with no blemishes.
- Fish stomachs and intestines obviously carry the same functions as in a land-based animal: they hold the consumed food and waste of the fish. It is important that the contents of the stomach are not in any way omitting foul odours or causing damage to the organ itself; it would need to be discarded if so.

Extraction

The stomach and intestine of a fish can best be removed from the cavity of a fish by gutting the fish conventionally in one piece with little mess (see page 49). Once removed, set the fish aside and position the organs on the cutting board as they were in the fish. The stomach looks like a small mitten sitting at the base of the gills, and the intestine can be fairly easily identified. Using sharp scissors or a small knife, cut the gills off the main body of organs, then snip all the other organs away from the stomach. Set aside the stomach and sort the intestines. This is best done by wiping away any mucus or excess fat from around the organ with kitchen paper.

Being careful not to tear the intestine, squeeze out the contents and discard. Do the same to the stomach by inverting the organ and scraping it clear with a spoon. Once the contents are removed, proceed to wash the stomach and intestines under cold water and then place on a clean stainless steel tray.

Application

- Given a fish only has one stomach and a single long intestine, this application needs to be somewhat of an accumulated method. So each time a fish is processed, remove these organs and clean them according to the extraction method and then bury them in fine table salt. Make sure the salt completely covers and fills these organs as it will not only draw moisture but remove any volatile aromas or potential bitterness that they may have. Salt for at least a week depending on size until completely cured. Be sure to rotate new stomachs and intestines going into salt with older ones coming out.
- Once cured, the salted organs can then be vacuum-sealed and frozen until required. If using straight away, the stomachs and intestines will need to be soaked in cold water for anywhere between 12 and 24 hours, depending on the size of the offal. For offal that has sat in salt for weeks or even months, this process of reducing the salinity by soaking in cold water will take longer.
- To cook the stomachs, the simplest method is to steam them until tender – this can be done in an oven or on the stove. Once tender, they can be used just like a beef tendon or pig's trotter would. Add them to a fish pie, a tomato-based ragu for pasta dishes, or simply fry or pickle them to serve on toast. The sky is the limit in terms of the opportunity, and the best advice is to see them as a vehicle to carry big flavours. Stomachs can also be sliced into rounds, sautéed and braised on the stove from raw until tender.
- The above methods also work well for the intestines, as does crumbing and frying or cooking and assembling the intestines as if they were pasta and building the sauce around them. You can also dice them very finely and include them in vinaigrette. Again, the creative opportunities are endless – what's critical here is retaining the intestine's tender texture and not serving it tough (or too soft if cooked too far).

TAILS

I've included tails with offal here as, in much the same way, they are too often tossed to one side for more desirable parts of the fillet. The problem is, they're very rarely used thoughtfully, with the tail cuts in restaurants often left to sit idle until no amount of lemon juice can save them as no-one wants to be serving up (or served) the end when everyone else gets handed the centre-cut brick out of the middle.

This is a terrible waste of what is a potentially magical part of the fish, and one that can provide us with so much critical information. One of my best uses for the tail early on at Saint Peter was to cut it off and cook it, then examine the texture to check for parasites that can render the flesh completely mushy (kudoa thyrisites) or tough as a board (tough fish syndrome) and get an understanding of what that fish would taste like. The cooked tail can also give an insight into how the fish was caught, killed and transported to us and if the cold-chain handling was managed appropriately.[5] When discovering and implementing my dry-ageing program before and after opening Saint Peter, it was also essential that we cut, cooked, ate and made decisions based on the textures and tastes of a day 17 fish and note how this differed to our day 4 experience, and cooking the tail in this way would enable us to do this.

Quality Points

The quality points for the tail of a fish are consistent with the quality points of the whole fish. You are looking for a nice coating of slime on the scale, a well-formed structure of scales that is unmarked or undamaged, no odour and glassy, firm flesh that is not discoloured in any way.

Extraction

The tail can be simply cut off the fish at the height at which you do not wish to serve and sell your primary fillet. In some cases, such as our Cold-smoked Tuna Tail Ham on the Bone on page 182, we cut the tail quite high as the sinews on a tuna run higher than on most conventional round fish.

Application

As well as being a vital informational tool as described above, boneless tail flesh that has been filleted and removed can be cooked and turned into great fishcakes, fish fingers or croquettes, blended with egg whites to produce a beautiful consommé or turned into a dumpling filling. Will the demand be there though? And will a chef just wanting to serve a centre-cut fillet of fish on the plate without any compromise be willing to introduce the 'waste' to the plate? Let's hope that the fish shops of the future can do the heavy lifting and see to it that fishcakes, fish fingers, fish pies and even fish and chips utilise this delicious section of the fish.

5. If a tail started dispelling water into the pan during cooking, for instance, we knew that the fish was more than likely killed and not ice-slurried correctly, meaning then that the flesh was already 'cooked' from not just the heat but the retained lactic acid residing within the flesh.

FAT

Fat is an incredible part of a fish to work with when well-harnessed. As there is often little to no visible visceral fat within the cavity of a fish, this component is not going to be hugely familiar as an extracted item. But it is seen in both wild and farmed fish, so for that reason I felt it important to include in this section.

Fat is not just limited to what resides within the cavity, however, and a high-quality, seasonal, well-fed fish will also carry a nice healthy coating of fat just beneath the skin. Too often the skin is unnecessarily taken off prior to cooking and with it goes the fat – a missed opportunity in my eyes as whether steamed, poached or baked, the skin can easily be peeled back after cooking. During cooking this fat melts down over the fish, moistening the flesh and imparting its own unique flavour profile.

Quality Points

- Fat can vary in its colouring, ranging from a tinted yellow through to white, off-white, cream and even orange within salmonoid species. Like many other organs set inside the cavity of the fish, the fat can also be compromised due mainly to the gall bladder being punctured and its bile causing staining and imparting a very bitter aftertaste, which makes careful extraction essential.
- The fat should be as odourless as the fillets. Its texture, again subject to seasonality and species, will vary from a thin webbing that surrounds the fish organs and almost looks like caul fat from a pig through to small soft greasy pieces hidden in and around the liver and intestines, and the large firm white blocks of fat extracted from aquaculture species of fish.

Extraction

To remove the fat of the fish, once the cavity has been opened and all the organs and gills have been set aside onto a board, take a small knife and cut the fat away from the intestine and stomach, being careful not to puncture the gall bladder or any other organs. In other cases where the fat is finer and less prominent, it can be simply pulled away with your hands.

Extraction can be seen as a physical action of removing something, but in the case of fats it can also be thought of as unlocking the potential flavours within the fish. A critical part of the dry-ageing process is a controlled reduction in moisture with an associated promotion of the fats that naturally reside within the flesh of the fish. This fat carries much of the unique identity of the fish and makes it simpler for the cook to decide on flavours to pair with it, rather than the stereotypical wedge of lemon that neutralises or suppresses a lot of what could've been. The challenge, however, in balancing the right amount of moisture with the right amount of fat within a fish when dry-ageing is not pushing it to the point where the fish is too dry and there is too much fat present, as this will result in the fat going rancid and sour.

Application

- There are a multitude of applications for the physically removed blocks of visceral fat within a fish, ranging from curing it into a lardo (see page 250), rendering it into a liquid fat for cooking (see page 246), using it within alcohol-based preparations (see page 258) or making soaps and candles (see pages 261–2). The sky really is the limit!
- Similarly, with regards to the application of fat that resides within the rest of the fish, all methods of cookery are on the table. Grilling a fish over coals, where the fat slowly drips down causing a wisp of smoke to hit the skin and enter into the pores of the flesh, results in a wonderfully smoky outcome. Poaching, steaming or pot-roasting a piece of fish with the skin on and bone in will give a beautiful texture from the gelatinous quality of the skin and bones, while the fat trapped under the skin will also be there waiting for you to peel back and enjoy.[6]
- The correct selection of species and method is also an important factor when trying to showcase the fat of a fish. Selecting a rich, fatty fish and battering it is not a good marriage, as there is no harmony or balance for the consumer. A fish that carries little fat and is naturally very lean also does not perform well when baked in an oven, as this eliminates the opportunity of experiencing the true taste and texture of the fish.

6. Another wonderful method of cookery to coax the natural fat from within a fish to the surface is to very carefully remove the skin from a fish, being sure to leave behind the fine layer of fat underneath. Set the skin aside for another application. Season the flesh of the fish with a little flaked or coarse sea salt, then take a hot piece of charcoal with a pair of tongs and carefully push the coal against the fish, only keeping it there for a second. Move the coal along the rest of the flesh, only ever holding it down for a second. This burst of extreme heat does two things: firstly, it gives the fish an incredible flavour that penetrates the whole fillet, and secondly, the severity of the heat from the coal creates a very thin layer of damage on the flesh, resulting in an uncompromised texture (as opposed to the preparation of tataki in a frying pan, using a blowtorch or grilling skinless fish directly on a grill, where the flesh toughens and dries too much).

BONES, CARTILAGE AND FINS

The bones, cartilage and fins of a fish are similar in all species. Some people may not have any issues with the wastage of these particular parts as, once gutted and scaled, the fish may be sold whole or as a bone-in cutlet. Irrespective of this, I felt it important to include these parts of a fish here as they really aren't considered enough outside the realm of making stock.

Extraction

The extraction of these items is subject somewhat to how you prepare the rest of the fish. Scissors, knives and fish pliers are all important tools when filleting or processing whole fish. Knowing the anatomical composition of where bones sit allows you to be a little more intentional about what gets cut first and what will be left at the end.

Road-mapping the cutting before the knife enters the fish will bring good solutions for the final product and also help with the length of time you are able to store a fish for as once the fish comes off the bone, the clock starts ticking. The bone plays a critical role in shielding the flesh from exposure to oxygen and light, which can alter the taste and texture of a fish rapidly and dramatically.

Application

In terms of how to use these items from a fish, the easiest way is to cook a fish on the bone so that you allow the flesh of the fish to benefit from the extra flavour and gelatine that resides within the bone. Also, by cooking on the bone, the fish holds its shape and moisture far better, and there is a little bit more wiggle room when it comes to timing the cook due to the protection the bone provides.

Many would say they prefer to eat a fillet than a whole or bone-in piece of fish. One reason for this is a fear of the smaller rib and pin bones of a fish that may have either caused us or someone we have seen to choke. The way to combat this is to leave the fish whole on the spine bone but cut out the rib bones and pull out the pin bones. Yes, this requires more time and effort, but it may be just the extra step needed to give someone a memorable eating experience.

With regards to bone marrow, as long as the fish is incredibly fresh and you are aware of where it is from and that it has been dry-handled, this delicious jellied morsel is worth the effort of cooking. I have found the marrow of tuna, swordfish and grouper all outstanding when roasted whole or even poached alongside a portion of the fish that it has come from. There are a number of ways to enjoy it, including smooshed into toast with herbs and pickles or simply seasoned with salt and pepper and served with tender potatoes.

If there isn't an immediate use for the fins of the fish then, once they are scissored off, they are easily frozen until enough are accumulated. The addition of these fins to a fish stock or soup is brilliant as they carry so much gelatine. The reason gelatine is important (and not just in fish cookery but in all types of cuisine) is that it acts as a conduit for flavour. Gelatine, like fat, will also increase the length of taste that something has. Natural gelatines found within the trimmings of a fish will also add texture and viscosity to dishes that may otherwise have lacked richness. Sauces or dressings can be made from the bones to elevate the taste of the fish that they were from or be used as a stand-alone product that dresses vegetables or leaves. Outside of gastronomy, the bones of a fish also carry the potential to be made into bone china or a bone glaze for ceramic plateware (see page 265).

MAW

The maw, swim bladder, bubble or float of a fish is found behind the organs in the fish closest to the spine. There are a number of techniques you can apply to utilise the maw, however note that as there is only one per fish, it is important to preserve the maw until you have enough to work with.

Extraction

The maw is best removed from the cavity of a fish by gutting the fish conventionally in one piece with little mess (see page 49). Once the organs are removed, look for a large blown-up balloon shape, though it may have popped and be hidden behind the offal. Cut the maw away with a sharp pair of scissors. It is important that the maw is unblemished and free of any odour.

Application

To preserve the maw, start by cutting it open as if you were cleaning a hood of calamari, then use the same knife to scrape away any residue or imperfections so that it is one even texture. On a steamer tray or steaming basket, steam the maw until soft. This can take anywhere between 10 and 35 minutes, depending on the fish. Once soft, remove from the steamer and transfer on a tray to an oven set to as low as it can go, or use a dehydrator to ensure that the maw dries out completely, approximately 12 hours.

Once thoroughly dried, maws are ready to be used in a multitude of applications. They can be stored dry in a clean container with a fitted lid or alternatively cryovacked to keep thoroughly dry. The maw can be rehydrated in water, stock or fragrant liquid for use within soups, stocks and sauce or, alternatively, could be deep-fried from dehydrated until crunchy and tripled in size. Ensure you fry in hot oil set to 200°C (390°F) as this intense heat will puff the maw dramatically and make a great textural inclusion to raw fish dishes or ground to season grilled or roasted vegetables.

THE CUTS

BASIC FILLETING

Be sure to start with a gutted fish with trimmed fins. For the first cut, position the fish with the belly facing you and the head to the left (or the right if you are left-handed).

1. Pull the pectoral fins outwards and make a cut behind to separate them from the fillet, then cut around behind the head until you hit bone. By doing this you are effectively separating the fish collars from the fillet.

2. Turn the fish so the belly is facing away from you (head on the right, tail on the left), then, starting from the cut at the top of the head, cut along the backbone from the head to the tail, cutting smoothly along the length of the fillet. Angling your knife towards the bones, keep running it along where the flesh meets the bones to open out the fillet until you feel it reach the raised spine in the middle. Using your knife, stay as close to the spine as possible and go over the bone.

3. Place the knife flat against the backbone and push the point through to the other side of the fillet. With the knife protruding out the other side and pressing against the spine, cut all the way to the tail to separate the tail section.

4. Lift the tail section to expose the ribs. Snip through the ribs with kitchen scissors up to the first cut. You can now remove the first fillet.

5. Flip the fish so the belly faces towards you and the head points left. Repeat the first cut, then cut along the back through the rib bones and, guiding the knife by pressing it against the ribs, cut towards the pin bones, then turn the knife the other way and cut up and against the ribs, using the bones as a guide and gently peeling away and slicing as you go.

6. Cut the second fillet away from the frame using scissors and wipe clean with paper towel.

FILLETING (HEAD AND TAIL ON)

Be sure to start with a gutted fish with trimmed fins, with the belly facing you and the head to the left (or to the right if you are left-handed).

1. Starting from the very top of the head, cut smoothly along the backbone from the head to the tail. Angling your knife towards the bone, run it along where the flesh meets the bone to open out the fillet until you feel your knife reach the raised spine in the middle, then place the knife flat against the backbone and push the point through to the other side of the fillet. (It is critical here that the knife stays as close to the bone as possible to prevent any flesh from being left behind around this tail section.) With the knife protruding out the other side and pressing against the spine, cut all the way to the tail being sure to slowly split the tail in half, allowing a tail presentation for both fillets.

2. Lift the tail section to expose the open cavity where the offal once was and where the ribs are connected to the spine. At this point, the ribs and the head are the only two sections that are keeping the fillet in place.

3. Snip through the ribs with kitchen scissors all the way up to the head. The fillet should now be separate from the spine.

4. Turn the fish so the head is immediately in front of you and tail furthest away. Have the fish cavity side up and turn the cut fillet flesh side up to the left. Using a sharp cleaver, cut down the left side of the spine and split the head in half. By doing this the fillet will now be completely free. Set aside the first head and tail intact half fillet.

5. Flip the fish so the belly faces away from you and the head points left. Follow the same process as for the initial fillet, starting this time by putting the knife in at quite an acute angle at the tail end, gradually lowering the angle and guiding the blade towards the head along the backbone, maintaining close proximity to the spine. Once you have reached the centre of the fillet, cut the tail section of the fillet away from the bone, ensure you take the remaining half tail along with it for presentation purposes.

6. Use scissors once more to cut the ribs off the spine and, once at the head of the fish, use scissors to cut away the dense bone from behind the last half of the head. (If the scissors can't get through this bone, use the cleaver to cut away.) Using a small sharp knife, cut away the pin bones from the ribs, then sweep the blade under the ribs. Repeat on the other fillet, then use pliers to remove the pin bones, being careful to simply wipe the pliers on a piece of kitchen paper next to you instead of dipping them into water each time (as even wet pliers here will be detrimental to the longevity of the fillet).

COLLAR AND BELLY 'SUPREME'

This cut harnesses the darker, fattier meat that lies within the collar of the fish alongside the intramuscular fat-rich belly, and it is absolutely delicious. The idea to remove these two cuts together and serve them as one piece came from a chicken supreme, where the wing tip is left in place on the breast of a chicken. To achieve this cut, start with a traditionally gutted fish. If you are right-handed, position the fish with the head on your left and tail to the right.

1. To achieve this cut, start with a traditionally gutted fish. If you are right-handed, position the fish with the head on your left and tail to the right, then, using a short sharp knife, cut from the anal vent of the fish in a straight line back towards you until it reaches the top of the fins closest to you. Turn the fish over so that the cavity is still away from you but now the tail is on your left and head on the right. Repeat the exact same cut again to ensure that the two cuts you have made marry up in a straight line at the top of the fish.

2. Insert a short sharp knife into the flesh immediately behind the head of the fish so your knife touches the hard collar bone on the left side of the blade. Draw the blade down from the very top point of the fillet behind the head at an angle that is consistent with the collar bone, being sure to stop a quarter of the way down so you don't cut the collar off the belly. Once you have reached the seam in the muscle that separates the mid loin of the fish from the belly, cut to follow this seam down to the anal vent.

3. The position of this seam sits just below the end of the rib bones, so it will be completely boneless except for the collar at the end. This cut will now look like a chicken breast shape. It will still be attached at the top of the collar, which can be easily snipped away using scissors. The final cut should now have the collar intact in one piece with the full length of the belly attached.

BUTTERFLY (HEAD ON)

This method can also be done with the head off, but when there is so much good meat on the head, why would you want to remove it?

1. Assuming you're right-handed (otherwise reverse these directions), place a whole scaled fish on a chopping board with its head to your left and tail to your right, with the belly facing you. Using a short sharp knife, draw the tip down the backbone of the fish from the head end to the tail, cutting along one side of the bone. Angling your knife towards the bone, continue to run it along where the flesh meets the bone to open out the fillet until you feel your knife has reached the raised spine in the middle. Place the knife flat against the backbone and push the blade across so that it stops just short of exiting the fillet. (The knife must stay as close to the bone as possible to prevent any flesh being left behind around this tail section.)

2. Open the tail section you have just cut into to expose the beginning of the cavity containing the offal and where the ribs are connected to the spine. Snip through the ribs with kitchen scissors all the way up to the head.

3. Once the fillet is free from the connecting ribs but still held in place by the tail, head and belly, snip the visible gills from just beneath the head. Using the heel of a chef's knife, butterfly the head through the back so that it flattens the head rather than cuts it in half.

4. Pulling from the freed gills towards the tail end of the fish, lift out the offal in one piece. Draw your knife down the backbone again for the second fillet, however this time the knife's first entry point is at the tail end and is drawn up towards the head end of the fish. Once the fillet is free from the spine but the skin of the fish is still fully intact, proceed once more with scissors to cut the ribs off the spine. Continue using scissors to snip out the backbone by cutting it just behind the head and just in front of the tail to give a kite-shaped fish with the tail intact and a butterflied head. Use fish tweezers to remove pin bones and rib bones. (Alternatively, depending on species, it may be easier to remove the rib bones with a small sharp knife.)

REVERSE BUTTERFLY (HEAD ON)

Make sure the fish is scaled and gutted conventionally before attempting this method. Position the fish with the head nearest to you and the tail furthest away.

1. Using sharp kitchen scissors, begin cutting down the left-hand side of the fish spine to disconnect the ribs from the spine, but stop at the fish's anal vent. Repeat this step down the right-hand side of the spine. (This will give you a clear track to use your knife in the next step.)

2. Position the fish now with the head on your left and the tail on your right. Using a small sharp knife, draw the blade down the scissored opening that you have made next to the spine.

3. Repeat on the opposite side. When these two cuts meet up at the tail, use kitchen scissors to snip the tail, then just behind the head where it meets with the spine.

4. Pull the spine carefully off the skin of the fish, making sure to support the surrounding flesh so it doesn't rip or do damage. Use fish tweezers to remove pin bones and rib bones. (Alternatively, depending on species, it may be easier to remove the rib bones with a small sharp knife.) Using a heavy knife or cleaver, tap on the head to crack down through it and split the skull, spreading it out flat.

SHORT RIB OF TUNA

The notion of a short rib of tuna might seem a little far-fetched, but the title here comes from the aesthetic appearance of this cut rather than where it has been taken from. This particular cut is derived from the collar of the tuna, located just behind the head of the fish and sitting in line with the gills. As this is a good way to utilise a head of tuna, you can start here with the head only, collars intact.

1. Using sharp scissors, cut the top of the collar off the nape of the fish. Cut the pectoral fin and surrounding bone and flesh away from the collar, making sure to keep the cut straight and clean. Repeat with the remaining side.

2. Position the collar on a cutting board skin side up with the tuna's clavicle to your left. This will now create the longest edge of the 'short rib'. At the base of this, use a knife, scissors or cleaver to cut about 2 cm (1 in) above the pectoral fin, removing the flesh to showcase the bone clearly and create what looks like a handle of the 'short rib'. Once the flesh is removed, set aside for another application.

3. Once you have a prism of consolidated flesh attached to the clavicle of the tuna, turn this cut over and proceed to remove the skin from the muscle, being sure to retain as much flesh as possible.

4. Using your sharp knife, seek out the small and medium bones that sit in the centre of the collar and are simple to find with the blade of the knife. Just like meat butchery, the instruction is always to stay close to the bone, as this way, minimal meat will be cut away unnecessarily while the non-structural bones that inevitably hinder the enjoyment of this cut once cooked are removed. Finish the cut by scraping away any remaining meat and using paper towel to polish the end of the 'rib' bone. This cut is now ready to be grilled over hot coals or under a grill, or even roasted and glazed in the oven.

FRENCHED CUTLETS (RIB BONE IN, SPINE OUT)

This is a perfect cut for crumbing as, apart from the single rib bone, the frenched cutlets are completely boneless. To attempt this cut, be sure to start with a fish that has been gutted, scaled and is no less than 2 kg (4 lb 6 oz), as anything smaller will not have bones that are structurally strong enough.

1. Position the fish on the cutting board with the head to your left and tail to the right. Using a short sharp knife, cut from the anal vent of the fish in a straight line towards the collar, being careful not to cut through the rib bones that sit immediately below. Like a rack of lamb, the idea here is to remove the belly from half of the ribs to expose them. Work your knife up and around to remove the collar and belly, then turn the fish over so that the cavity is still away from you but now the tail is on your left and head on the right. Repeat the same cut again to ensure that the two cuts you have made marry up in a straight line at the top of the fish.

2. Using a cleaver or large chef's knife, slice through the soft vertebrae and then find the spinal vertebrae to cut in between, allowing you to then cut off the tail end of the fish, which can be set aside for another purpose. Draw the blade down in a diagonal line from the nape behind the head of the fish, past the pectoral fin and finishing on the tail side of the pelvic fin. Repeat this on the other side ensuring the two cuts marry up at the top, allowing you to cut the head off the fish. Set the head and collars aside for a separate application. What's remaining now is both fillets on the bone with the rib bones still in place.

3. Using a sharp pair of scissors, cut the ribs away from the central spine, being careful not to penetrate the full length of the scissors into the flesh, as this will compromise the final aesthetic. Once the ribs have been cut off both sides of the spine, cut down to seperate the fillets and remove the central spine. With a short sharp knife, cut the intercostals from between the ribs. This meat can be set aside for another recipe.

4. With a chef's knife now, cut between each frenched rib bone. This will create individualised cutlets and allow pin bones to be removed from the flesh. Repeat on the other side. To clean the flesh fully off these exposed rib bones, bring a pot of water to the boil and carefully dip the bones into the water. Once the bones have been briefly dipped into the hot water, use a tissue or towel to rub away the cooked flesh from the bone.

YELLOWFIN TUNA RIB EYE

This cut is a result of utilising the first loin and two boneless bellies for steaks or other preparations and then leaving the last loin on the spine and frame that it is attached to.

1. To start working with this last remaining loin of the tuna that is on the bone, use a sharp knife to cut away the cartilage and spines from the top of the fish.

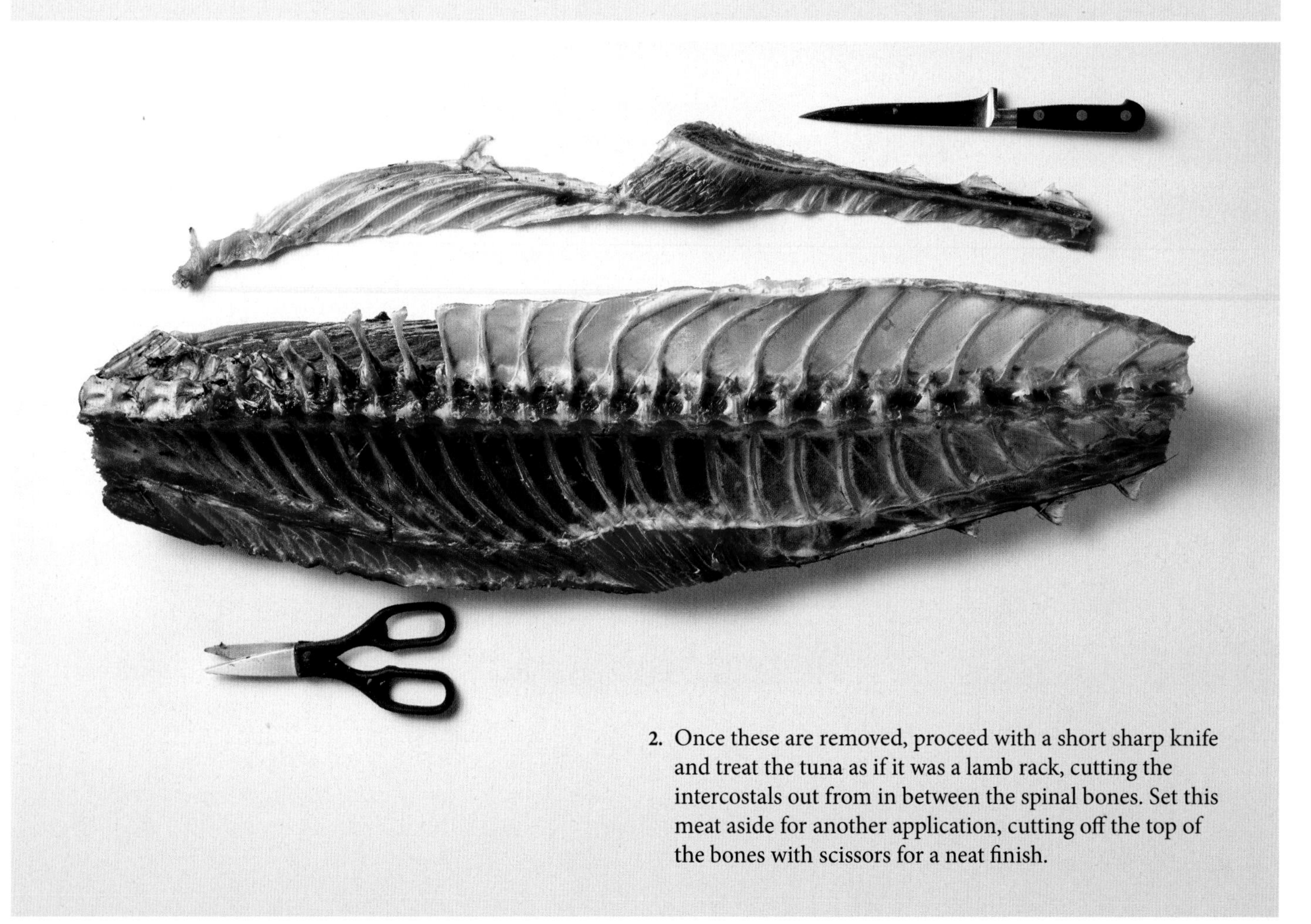

2. Once these are removed, proceed with a short sharp knife and treat the tuna as if it was a lamb rack, cutting the intercostals out from in between the spinal bones. Set this meat aside for another application, cutting off the top of the bones with scissors for a neat finish.

3. Once the bones are free of meat, they can be polished back to white with the assistance of some boiling hot water and paper towel.

4. The next cut is to remove the significant presence of the lateral swimming muscle off the side of the fish. (You don't want to take it all off as this carries a lot of the savoury qualities of the fish, however leaving too much in place can be a little overwhelming.) Start with a long sharp straight-edge knife and proceed to sculpt away the dark muscle from the side of the tuna, being sure not to take too much. This long cut of tuna is ready now to either be grilled or roasted in this large format or split on the vertebrae to separate the bones into single-bone steaks. It is then at your discretion to cook with the skin on and remove after cooking or take it off prior, which can be simply done with a sharp long knife, working as closely to the skin as possible.

BUTTERFLIED BELLY

This is a cut that I really enjoy cooking and one that I feel would be enjoyed by others if it became a common method of preparation. The cut must be made with a scaled but ungutted fish as the belly needs to be kept intact to execute it properly.

This large, boneless, triangular-shaped piece of fish is easy to grill over whichever medium you desire – I personally like to sit it on a wire rack and leave it to dry slightly overnight in a fan-forced refrigerator, as this only makes the skin crunchier when cooked over coals on a hot grill. If proceeding in this way, use a weight on top while cooking to create an even surface and start to cook the top of the fish to avoid the need to then flip the flesh onto the direct heat.

1. Select a short sharp knife and, starting just behind the pectoral fin and in line with the lateral line of the fish, make a cut that follows that line but begins to sweep back down to finish just behind the anal fin of the fish. Repeat this same cut on the other side of the fish to ensure both cuts meet up at the base of the fish.

2. Using your knife, cut the collar off where it meets the nape of the fish. Do this on both sides.

3. Scissor down the initial cut to separate the ribs from the spine. Using your knife, remove the fillet from the bone, leaving the cartilage intact at the tail end. Repeat on the opposite side.

4. Cut the belly off the fish to leave you with the whole belly in one triangular-shaped piece, setting aside the remainder of the fish for another application, such as Saddle chops (see pages 120–3). Use fish tweezers to remove pin bones and rib bones. (Alternatively, depending on species, it may be easier to remove the rib bones with a small sharp knife.)

PORTIONING A FISH FILLET FOR BATTERED FISH

Over the years of serving battered fish, we have encountered just about every known scenario of undercooked, overcooked, under-fried and raw batter or over-fried and burnt batter. Many of the issues stem from portioning and thickness, which are both critical to a memorable piece of fried fish – when portioning fish for battered fish, it's super important that the surface area is even and the size very similar from piece to piece.

What's also critical in terms of preparing a fish for battering is selection of fish species. To me the perfect fish for battering is one that has a little bit of density and thickness to it as this will assist in moisture retention and create a crunchy exterior and moist interior. It is also important to work with a fish that has a good amount of visible intramuscular fat – selecting a fish that is too lean or known to dry out when cooked past medium-rare will not perform any better in batter than in a frying pan or on a grill. It helps to see battering as not so much deep-frying but more an insulative steamed method of cookery that relies heavily on residual heat to continue cooking the fish after coming out of the fryer. And don't let nostalgia be the determining factor when selecting a fish for battering – doing so can restrict you from exploring a more diverse array of species and opportunities that might be better than what you ate growing up.

I prefer to leave the skin on when deep-frying as this will give the consumer a very clear understanding of the identity of the fish that you are frying from the point of view of flavour, as well as retaining more moisture and a good amount of gelatine. However, this decision needs to be informed by whether the fish you are cutting has skin that will be tender and soft once fried – there are a number of fish species we use where the skin must be removed as it just doesn't break down in the amount of time it takes for the fillet to cook.

It is completely at your discretion as to how thick or thin you slice your portions as fish will vary (as will batter recipes), however, bear in mind that thickness is critical in retaining a juicy interior while the batter has time to caramelise and develop that wonderful crunch. As a rough guide – and once you've determined whether the skin should be on or off – we tend to use the following method, selecting a long sharp knife to cut the portions as follows:

1. Place the fillet on a cutting board with the skin side down and the shoulder of the fish to your left and tail to the right.
2. At a 45-degree angle, make a slice approximately 4 cm (1½ in) from the shoulder of the fish through the flesh side down to the skin. Maintaining this angle and thickness, continue to slice portions through the centre of the fillet.
3. Once you reach the point of the fillet where it starts to taper away and become thin, increase the length between cuts to approximately 8 cm (3¼ in), creating a wider surface area but obviously a thinner cut due to the shallower depth. Portion these thinner cuts together to ensure cooking times are consistent.

SAINT
PETER

FISH PORTIONS FOR GRILLED FISH

In fish cookery, the cooking method, choice of species and decision of how to cut it to complement the method are all of equal importance for me – these three variables can fundamentally determine whether or not you will have a thoroughly enjoyable eating experience.

Grilling fish seems simple enough, however the majority of the time, due to the skin being wet, or the grill being too hot or not hot enough, we wind up in a bit of a mess. Having grilled fish over coals for the past six years at Saint Peter and then at Fish Butchery and Charcoal Fish, I find myself grilling more and more as it really is one of the most delicious and simplest ways to cook.

When grilling, I make sure I am selecting fish that isn't too thick. Working with a thick fillet of fish becomes incredibly difficult, even with the assistance of a fish weight, as the skin takes on colour disproportionately quicker to the time that it takes to cook the flesh. Furthermore, flipping the unprotected flesh side directly onto a hot grill will result in a compromised texture and aesthetic.

There is a long list of Australian species that I love to grill over coals, including rock flathead, garfish, flounder, King George whiting, mackerel, herring, bonito and sardines. The common thread between all of these species is that they are relatively small fish and the majority can either be cooked on the bone or butterflied. The beauty of cooking the fish whole on the bone is that none of the delicate flesh is exposed to the direct heat from the grill, meaning it has a better texture that isn't likely to resemble a dried-out piece of skinless chicken breast. Also, cooking on the bone immediately gives a lot of wiggle room in terms of cooking time along with imparting plenty of flavour from the bone and retaining more of its moisture.

Grilling a butterflied fish has to be one of my absolute favourite methods of cooking. Butterflying creates a generous surface area of skin. If the skin has been conditioned well through dry-handling and correct storage, you will have the opportunity to make all that skin incredibly crispy and smoky. As the skin crisps over the fire, the fat that resides beneath the skin escapes the pores and drips down onto the coals, creating a puff of smoke that then enters the crackled skin resulting in a wonderful flavour and texture.

If you want to grill a thick fillet on the barbecue then do so by cooking a whole fillet or at the very least a portion that exceeds one single portion. This will be far more enjoyable to cook than trying to juggle several individual portions that will require more maintenance and care than one single large piece. The best way to achieve a wonderfully crisp skin and a beautiful interior texture for a large piece of fish is to 'reverse cook' it in the oven. Baking the fish prior to grilling will give a beautifully even finish and allows you to carve it with ease after grilling.

To reverse cook for grilling, place the fillet on a wire rack in a preheated oven set to approximately 65°C (150°F) with a probe thermometer in the thickest part of the fish. When the thermometer reaches approximately 40°C (105°F), again depending on the species, remove it from the oven and leave it to rest for 5 minutes, then brush the skin with a neutral oil such as grapeseed and season liberally with salt flakes. Place this already cooked fillet of fish skin side down onto a wire rack over hot coals and place a weight on top. Move the weight across the top of the fillet to ensure an even amount of heat reaches all of the skin. Manage the heat by moving the rack on and off or placing the rack nearest or furthest away from the intense heat. Remove from the heat completely, flip the fish over to the flesh side while still on the rack and allow it to rest skin side up for 4–5 minutes, then transfer to a cutting board and carve the fish from the flesh side through to the skin.

SADDLE CHOPS (BONE IN, RIBS AND PINS OUT)

This cut is what I see as the 'best end' or saddle of fish. It can be left as one long piece or further cuts can be made from the flesh side of the fish to create chops from between the vertebrae. These individualised darnes or chops can be roasted, grilled or poached for one. Alternatively, roast this whole section of saddle or gently sauté and finish over hot coals to crisp up the skin.

1. Start with an ungutted and scaled fish here. Position the fish on the cutting board with the head to your left and tail to the right. Have the shoulder or dorsal side of the fish closest to you and the open cavity away. With your left hand, hold the belly of the fish to make it taut. Using a short sharp knife, cut from the anal vent of the fish in a straight line back towards you till it reaches the top of the fins closest. Turn the fish over so that the cavity is still away from you but now the tail is on your left and the head on the right. Make the exact same cut again, ensuring that the two cuts have married up in a straight line at the top of the fish.

2. Using your knife, cut the collar off where it meets the nape of the fish. Do this on both sides.

3. Scissor down the initial cut to separate the ribs from the spine. Using your knife, remove the fillet from the bone, leaving the cartilage intact at the tail end. Repeat on the opposite side. Cut the belly off the fish whole in one triangular-shaped piece and carefully remove the offal for other applications (see pages 57–77).

4. Using a cleaver or large chef's knife, cut through the soft vertebrae to cut off the tail end of the fish. This can be set aside for another purpose. Draw the blade down in a diagonal line from the nape behind the head of the fish, past the pectoral fin and then finishing on the tail side of the pelvic fin. Repeat this cut on the other side ensuring the two cuts marry up at the top, allowing you to simply cut the head and collars off the fish. Set these aside along with the tail for separate applications. You will now be left with both fillets on the bone, with the belly intact and the rib bones still in place.

5. Holding the saddle in your hand with the head end closest to your body, use a sharp pair of scissors to cut the ribs away from the central spine. Place the fish back down onto the cutting board and sweep out the rib bones with a short sharp knife. The cut can be left as one long piece or further cuts can be made from the flesh side of the fish to create chops from between the vertebrae.

STUFFED BONELESS SADDLE

This is about as complicated as you can get for a fish cutting and assembly technique. But then, this book *is* called *Fish Butchery* ... Think of it as the porchetta of the sea.

1. Poistion a scaled and gutted fish on a cutting board with the head nearest to you and the tail furthest away. Using a sharp heavy kitchen knife or cleaver, make an initial cut between the vertebrae on the spine and remove the head cleanly as a single unit. Remove the collars by making a further diagonal cut from the top of the shoulder of the fillet down to the belly of the fish just behind the first fin.

2. Using sharp kitchen scissors, begin cutting down the left-hand side of the fish spine to disconnect the ribs from the spine, but stop at the anal vent. Repeat this step down the right-hand side of the spine. This now gives you a clear track to use your knife in the next step.

3. Position the fish with the tail on the right. Using a small sharp knife, draw the blade down the scissored opening that you have made next to the spine. Repeat on the opposite side. When these two cuts meet up at the tail, use kitchen scissors to snip the tail.

4. Pull the spine carefully off the skin of the fish, making sure to support the surrounding flesh so it doesn't rip. Use fish tweezers to remove the pin bones and rib bones. (Alternatively, depending on the species, it may be easier to remove the rib bones with a small sharp knife by sweeping them out.)

5. Cut the top loins out of both sides of the fillet, being careful not to cut a hole in the skin. Set the top loin pieces aside as these will be put back in when it comes time to assemble the roll.

6. Next, place the blade of a short sharp knife into the thickest part of the fish, which is the mid loin that sits next to the vacant space you have created by removing the top loins. The entry of the blade should be on the pin bone side of the mid loin, not next to where the top loins have just come out. The aim is to make an even cut through this thick part of the fillet to create what looks like a book opening. The piece that butterflies open will fill the vacant space where the top loin once sat, so now you will have one flat, even layer of fish. The challenge in making this open-book cut is to not let the knife exit on the other side – you want the mid loin to still be connected so that it fits in the space snugly and doesn't want to slip out.

7. Similar to how the top loins were carefully taken off the skin without doing any damage, cut the bellies off the skin on both sides of the fish. Set the skinless whole bellies aside with the top loins in readiness to roll. What you should now be looking at is a rectangle of unpunctured skin where the bellies once sat, with the book-opened mid loins now spread across the space that was created by removing the top loins and a thin line of cartilage down the centre holding the two fillets together.

8. To create a nice finish, place a square of nori seaweed over this central square of flesh and push down so it adheres evenly. Wrap the two top loins of the fish together in another square of nori seaweed. Set these pieces aside in the refrigerator for a moment to come together.

9. On the central square of flesh, place one of the two bellies down from head to tail on the board. On top of this, place a sausage filling of fish trimmings (see chorizo on page 154) across the belly in one even layer approximately 1 cm (½ in) thick. Place the wrapped top loins along the centre of the sausage filling, running in the same direction as the belly. Cover the top of the rolled loins with a further 1 cm (½ in) thickness of sausage mix.

10. Place the remaining belly over the top of the sausage mix to cover the sausage meat completely.

11. Cover the belly and sides of the centrepiece with nori sheets and press in at the edges to sit tightly.

12. Now pick up the skin from the left side of the fish and pull the mid loin over to sit atop the centrepiece and repeat on the other side. Turn the whole cut over so that the seam is on the cutting board. Using kitchen twine, make a basic knot around the end closest to you, then loop the twine around your hand and fit it over the opposite end of the fish away from you. Move this loop down to a 2 cm (¾ in) gap from where the first knot was made, then proceed to loop twine all the way to the end. Tie off and chill for a further hour at least before roasting.

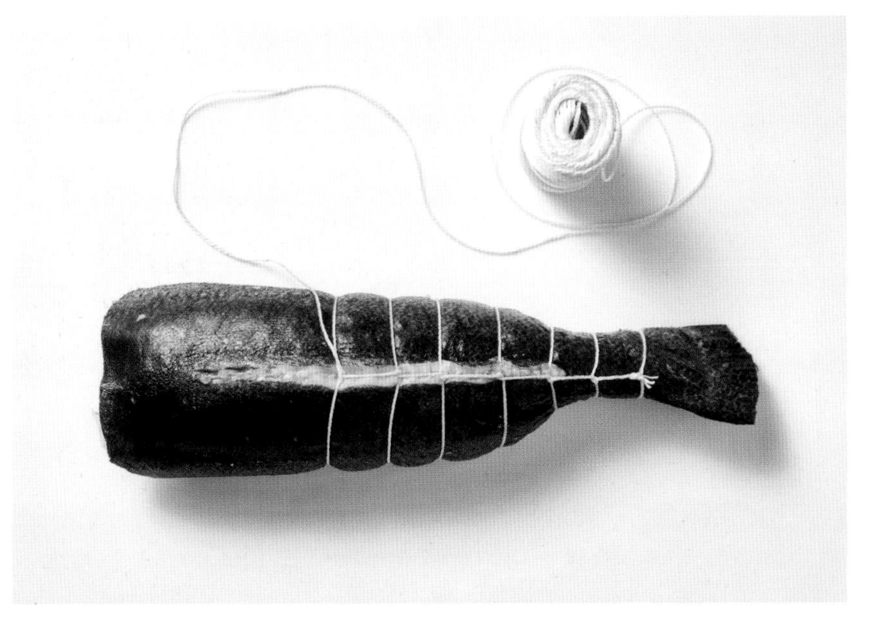

FISH CROWN STUFFED UNDER SKIN

This cut is inspired by a crown-roasted chicken. The butter adds a huge amount of flavour directly to the flesh while also keeping it moist. Further flavours can be added inside the cavity of the fish while it cooks. This is one of the more unconventional cuts in this book, but it's one that I couldn't leave out as the finished result is pretty special, especially if it's truffle season!

1. Before beginning, wash and sanitise a standard stainless steel knife-sharpening steel. Starting with a 4–5 kg (8 lb 13 oz) fish at minimum that has been traditionally gutted and scaled with the fins trimmed – and hung and matured for a minimum of 3 days to ensure the skin is nice and dry. Cut the tail off at the third vertebrae from the caudal fin.

2. Using a very sharp short knife, make a small incision between the skin and the flesh approximately 2 cm (¾ in) deep, working just beneath the skin, no deeper. Insert the sanitised steel into this incision and gently push it upwards toward the skin to avoid it puncturing the flesh. The aim is to separate the skin from the flesh all the way across the surface up to the top of the fish behind the collars. Pay close attention to where the belly is particularly thin. If the steel punctures or damages the flesh or your steel comes out through the skin, this will flaw your finished product. Once the skin is completely separate all the way around the fish, place in the refrigerator.

3. Prepare a compound butter, a softened salted butter with grated fresh black truffle or any other flavoured butter of your choice. Whip the softened butter and place in a piping (icing) bag. Remove the fish from the refrigerator, open the incision slightly at the tail and insert the piping bag of butter. Squeeze the butter into the incision and, with your other hand, work this butter all the way up to the collar and around the belly.

4. Once filled all the way around, return the fish to the refrigerator, this time on a wire rack sitting on its collars and tapping open the head, as shown. To cook, preheat the oven to 240°C (465°F) and roast the fish for a time proportionate to the skin becoming crisp and the flesh reaching an internal temperature of approximately 45°C (113°F). (This stuffed fish benefits from a higher temperature due to the added moisture of the butter, which you don't want to stew within the fish.)

DOUBLE SADDLE OR BARNSLEY (RIBS ON, PINS OUT)

The benefits of this cut are not only its visual beauty but the flavour gained by cooking on the bone and the retention of moisture and structure. This cut is perfect for grilling quickly over hot coals or poaching in a curry or stock.

1. Start with a fish that has been gutted and scaled, with fins trimmed. Position the fish on a cutting board with the head to your left and tail to the right. Using a cleaver or large chef's knife, cut through the soft vertebrae and then find the spinal vertebrae to cut in between, allowing you to then cut off the tail end (this can be set aside for another purpose). What you are left with now is the head end of the fish, which has one intact fillet on each side.

2. Position the fish back at the centre of the board now with the head on the left and cavity facing away from you. Draw the blade down in a diagonal line from the nape behind the head of the fish, past the pectoral fin and then finishing on the tail side of the pelvic fin. Repeat this cut on the other side, ensuring the two cuts marry up at the top, allowing you to then cut the head and collars off. Set the head and collars aside for a separate application. What's remaining now is both fillets on the bone with the belly still intact, meaning the rib bones are still in place.

3. Find the lateral line of the fish by looking for a visible line across the surface of the skin. Once found, depending on the size fish you are working with, place two fingers on the belly side of the line and make a small mark with your knife indicating where to cut. Using a long sharp knife, cut across this line being careful not to cut through the rib bones that sit immediately below (like a rack of lamb, the idea here is to remove the belly off half of the ribs to expose them). Repeat on the other side and set the bellies aside for another application.

4. Once half the ribs are exposed, proceed with a short sharp knife to cut the intercostals or the meat from between the ribs. This meat can be set aside for another recipe. Repeat on the other side. To clean the flesh fully off these exposed rib bones, bring a pot of water to the boil and carefully dip only the bones into the water, then use a tissue or towel to rub away the cooked flesh from the bone.

5. With a chef's knife now, cut each vertebrae within the open cavity of the fish leaving you with a portion of double saddle that has its frenched rib bones intact. Once cut, the couple of pin bones that will now be visible can be removed with fish pliers.

BONE IN FOREQUARTER (PINS OUT AND RIBS OFF)

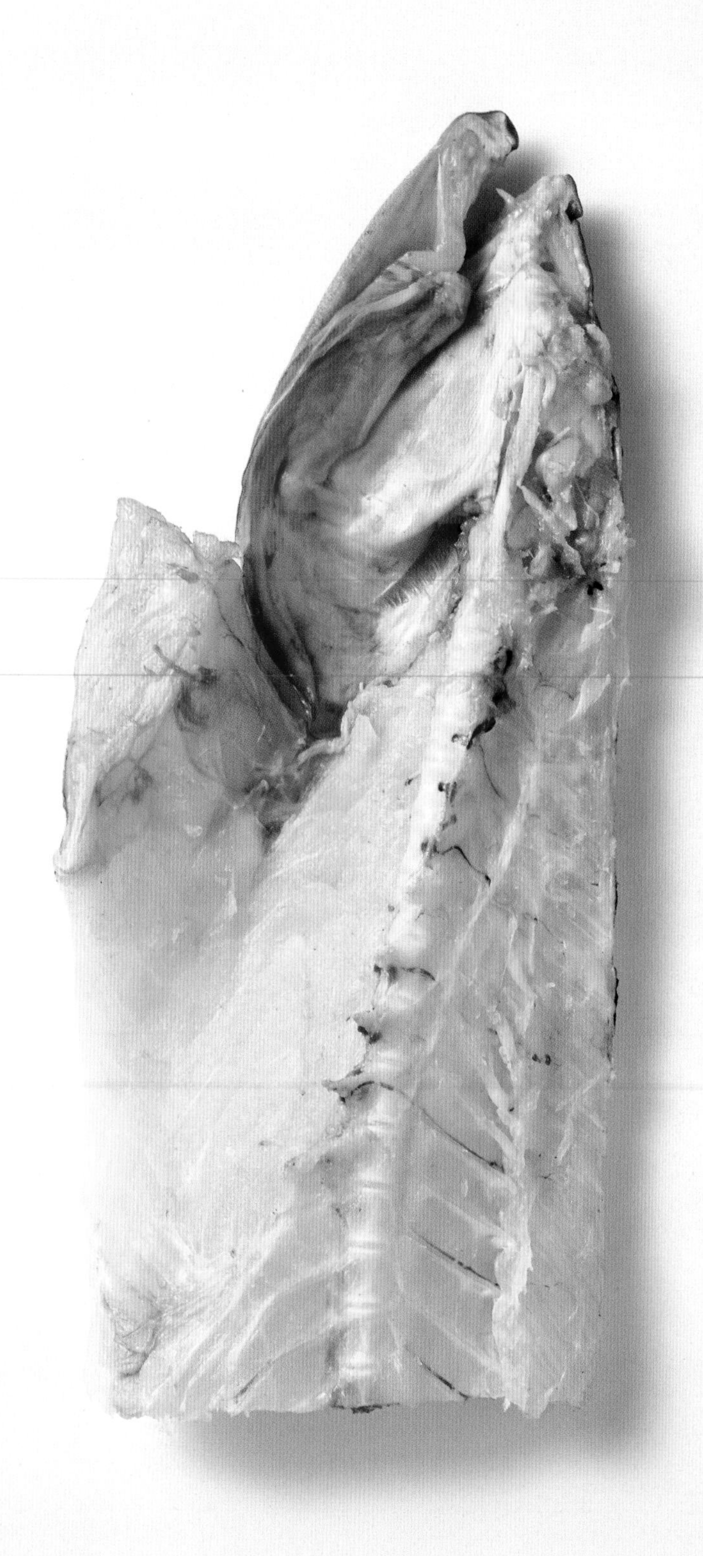

The final cut will have the complete spine and frame intact on the fillet but will be free of the inconvenience of the pin and rib bones. The beauty of this cut is that it can be grilled over coals with the freedom to cook it on both the skin and bone sides. This is such a delicious cut and one that looks very generous when it lands on the dinner table.

2. Cut the skin off the tuna then separate each vertebra into rounds.

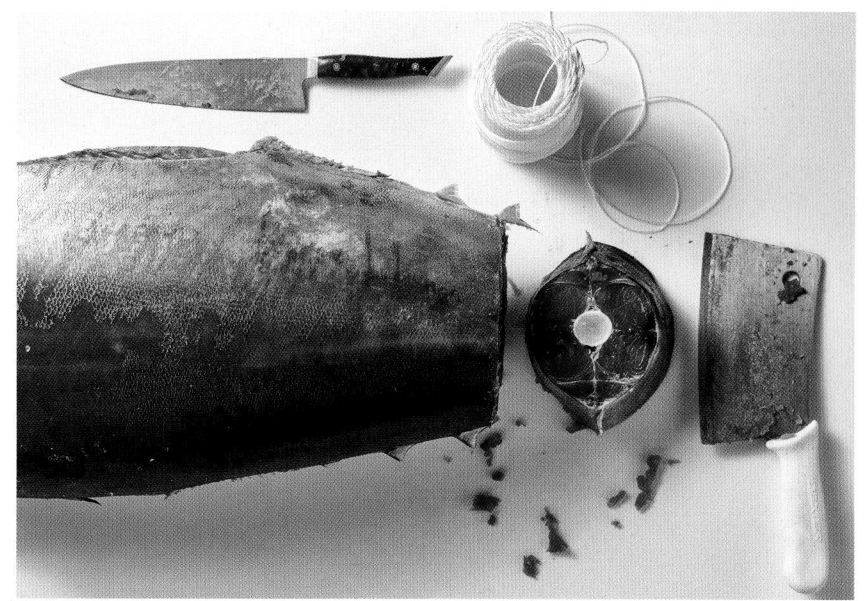

3. Once the rounds are cut, assemble the largest round in the centre of a cutting board and then position the smaller rounds around that central piece.

4. Using kitchen twine, make a double loop around the circumference of the rounds and tie them off firmly. Make a loop over your hand as if you are trussing the fish and place that around the diameter of the fish. Continue to add more loops on different tangents to help hold these rounds of tuna in place. When the tuna is firmly strung together, place in the refrigerator to set.

The tail rounds can obviously be cooked independently, but to make the most of holding them together like this, the strung tuna round can first be marinated in wine and aromatics like thyme, rosemary, juniper and orange zest. Dust the whole marinated round in a little plain (all-purpose) flour to assist in thickening the sauce around it in the end, then brown the tuna off in a hot frying pan until evenly tanned. Remove and drain well, then add the marinating wine to the pan and deglaze. Reduce the wine to a glaze and add a dark fish stock on top of this, or follow the recipe for Fish Jus (see page 245) and use this as the basis of a sauce. Reduce by half and then add the whole round to the pan to complete the cooking of the tuna. Unlike oxtail or beef cheeks, this tuna will not take long to cook. The final product should still look medium and nice and rosy, but the benefit of working with the tuna tail is that the spine is full of gelatine and marrow. Once it softens down, this marrow and connective tissue enriches the sauce that the tuna is cooked in and gives you something completely unique and a dish that will wow everyone. This is best served with a good quality mashed potato or, alternatively, the wine-based braise can be substituted for a braise of tomatoes and peppers and then served with a fresh corn polenta.

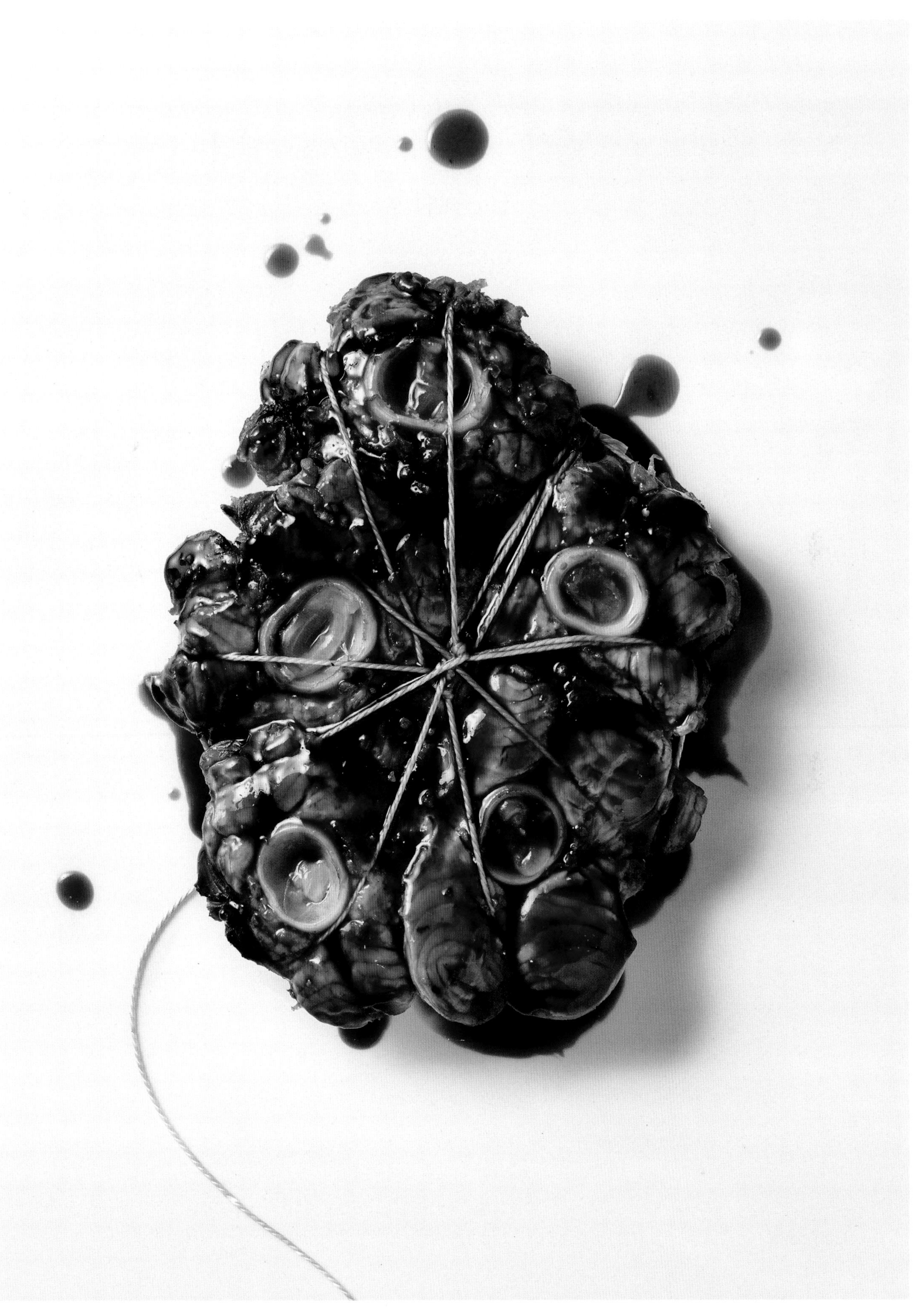

CRAFT

A butcher's greatest opportunities lie within the sundries of a single animal. We talk about the significant ethical responsibility in great detail in this book, but what's as important is the economical responsibility of the business owner. Great margins can be achieved by creating value-added foods with these parts. If butchers never applied themselves to this aspect of the craft then we wouldn't have burgers, hot dogs, ham, sausages or salami.

This section of the book highlights the many creative opportunities we are missing, not just from a gastronomic standpoint but from extensions such as fish-fat soap, using fish bone in ceramics and the utilisation of fish offal in alcohol production.

Deriving such craft items from one fish is part of the way we must see fish in the future. And it is the industry's responsibility to bring this to action.

CHARCUTERIE

FRESH CHORIZO

What's most important about this fish chorizo recipe is the 2 days of hanging after making. Hanging these sausages allows the casings to dry slightly, making them far simpler to cook. Once the cases are nice and dry after 2 days, the chorizo can be covered and stored or, alternatively, frozen.

MAKES 9.4 KG (20 LB 12 OZ) OR ABOUT 74 SAUSAGES

4 kg (8 lb 13 oz) fresh boneless, skinless Murray cod (or gurnard)
3.5 kg (7 lb 12 oz) fresh boneless, skinless salmon
1.5 kg (3 lb 5 oz) Cured Fish Fat (page 250)
10 garlic cloves, microplaned
30 g (1 oz) cure #1
500 g (1 lb 2 oz) quinoa flakes
120 g (4½ oz) table salt
80 ml (2½ fl oz/⅓ cup) white vinegar
60 g (2 oz) chilli flakes
15 g (½ oz) finely grated nutmeg
15 g (½ oz) dried oregano
15 g (½ oz) ground cumin
32 g (1 oz) ground black pepper
40 g (1½ oz) garlic powder
200 g (7 oz) smoked paprika
200 g (7 oz) sweet paprika
ox casings

Place the Murray cod, salmon and fish fat in the freezer. Once the fish is below 0°C (30°F), mince through a 1 cm (½ in) plate, then combine with the remaining ingredients and mix very well by hand (a mixer can overwork the mixture) until sticky, approximately 10–15 minutes. Place the mixture in the refrigerator for a minimum of 2 hours.

Soak casings in water for 1 hour prior to filling.

Set up a sausage filler. Temperature check the mixture during mincing and filling, aiming to stay below 8°C (45°F). When filling the casings, make sure the work surface is damp, so the sausages don't stick. This can be done with a spray bottle of water to mist over the bench.

Poke the sausages with a needle if you see any obvious air pockets and link the sausages by twisting them at 12 cm (4¾ in) intervals.

Hang sausages on butcher's hooks in a conventional refrigerator set between 1° and 2°C (33 and 35°F). If keeping longer than 2 days, cover the sausages to prevent overdrying.

These fresh chorizo can be simply pan-fried in a little oil or alternatively grilled over coals until just done. Subjecting the sausages to too high a heat can cause the skins to break. If cooked for too long, you run the risk of drying out the sausage as there is naturally less fat in this sausage than a conventional meat-based variety.

TUNA CHORIZO CASTELLANO

Unlike the Fresh Chorizo that also appears in the book (page 154), this is a cured and dried chorizo that can be sliced and consumed as a cold cut, similar to a salami.

MAKES 4.5 KG (9 LB 15 OZ) OR ABOUT 25 SAUSAGES

- 5 kg (11 lb) boneless, skinless tuna trim
- 2 kg (4 lb 6 oz) boneless, skinless cold-smoked salmon fillet
- 500 g (1 lb 2 oz) Cured Fish Fat (page 250)
- 100 g (3½ oz) table salt
- 18 g (¾ oz) cure #2
- 15 g (½ oz) garlic powder
- 15 g (½ oz) onion powder
- 30 g (1 oz) ground black pepper
- 15 g (½ oz) ground white pepper
- 7 g (⅛ oz) ground dried oregano
- 10 g (¼ oz) ground nutmeg
- 40 g (1½ oz) paprika
- 10 g (¼ oz) sodium erythorbate
- ox casings

Place the tuna, smoked salmon and fish fat in the freezer. Set up a commercial meat grinder fitted with a 13 mm (½ in) plate.

Once the fish is below 0°C (30°F), mince through the grinder. Mix together with all other filling ingredients, then place the sausage mixture in the coolroom for at least 1 hour until ready to fill the casings.

Soak casings in water for 1 hour prior to filling.

Set up a sausage filler. Temperature check the mixture during mincing and filling, aiming to stay below 8°C (45°F). When filling the casings, make sure the work surface is damp so the sausages don't stick. This can be done with a spray bottle of water to mist over the bench.

Using cotton butcher's twine, tie the sausages at intervals about 15 cm (6 in) long, making sure they are filled tight to avoid separation. Poke the sausages all over with a needle to allow moisture to escape; you want a weight loss of about 30 per cent. Hang the sausages on butcher's hooks and store in a well-ventilated refrigerator to dry. (These sausages are not fermented, only cured, so a salami cabinet is not necessary.) Allow 4 weeks and slice open to try – depending on the humidity of your fridge, this process may take less or more time. The desired result is one that is slightly chewy, not dry and crumbly, nor too soft in the centre. Once this desired texture is achieved, store in cryovac to avoid the risk of the salami drying out further. This can be simply served as a cold cut.

MORTADELLA

This recipe has become a week-to-week essential for us at Fish Butchery as it utilises the trim from both white fish and smoked salmon. The smoked salmon trim is used here as a way of getting a beautiful, well-rounded smokiness that doesn't rely upon smoking the mortadella itself. Once cooked and chilled, the mortadella can simply be sliced on a meat slicer or even blended into a mousse to stuff inside filled pasta like agnolotti or ravioli.

This mortadella recipe has been refined and tinkered with over the past few years. I can finally now say that it is absolutely delicious.

MAKES 2.7 KG (6 LB) OR
2 MORTADELLA SAUSAGES

350 ml (12 fl oz) water
150 ml (5 fl oz) fish sauce
1 kg (2 lb 3 oz) skinless, boneless white fish trim (snapper, flathead, cod, gurnard, mullet)
1 kg (2 lb 3 oz) skinless, boneless smoked salmon trim (alternatively, salmon tails and trim can be used)
10 g (¼ oz) table salt
8 g (¼ oz) caster (superfine) sugar
250 g (9 oz) egg whites
20 g (¾ oz) coarsely cracked black pepper
500 g (1 lb 2 oz) cubed Cured Fish Fat (page 250) or Fish Bacon cubes (page 174)
75 mm (3 in) plastic casings

Start by setting up a commercial meat grinder on a clean work surface and placing a Robot-Coupe processor jug and blade in the freezer.

Before starting the mincing, combine the water and fish sauce in a measuring jug and place in the freezer. Leave in the freezer until the liquid is completely chilled but not frozen.

Mince the white fish and smoked salmon trim through a 2 mm (⅛ in) mesh plate into a clean stainless steel bowl and then put in the freezer until very cold.

Remove the minced fish from the freezer and place a quarter of the mix into the cold Robot-Coupe jug. Blend the fish mince with a quarter of each of the salt, sugar, egg whites and cold water and fish sauce solution, then blend till very smooth. Transfer to a cold stainless steel mixing bowl and repeat with the three remaining quarters.

Using a rubber spatula, fold through the coarsely cracked black pepper and cubes of cured fish fat and place the mix in the refrigerator until ready to fill the casings.

Using a sausage filler, fill two 75 mm (3 in) plastic casings with the mortadella mix. Secure both ends of the casing very tightly with butcher's twine.

Hang in a combination steam and convection oven on butcher's hooks. This is done so that the finished product stays perfectly cylindrical and cooks very evenly. Steam at 70°C (160°F) for 30–35 minutes, or until the internal temperature reaches 58°C (135°F) on a digital probe thermometer. Allow to cool for an hour before refrigerating. The mortadella can now be sliced on a sharp meat slicer and served as is.

TUNA PISTACHIO SALAMI

Fish salami? Although the casing that I have used here is from a cow, this otherwise completely fish-based salami is not only extremely delicious but also remarkably similar to a meat-based product. This recipe was developed to utilise the scraps of tuna that we would accumulate from breaking down whole fish. In many ways it seemed startlingly obvious to reach this outcome – where would a butcher be without a recipe that solves the issue of animal sundries?

MAKES 4.5 KG (9 LB 15 OZ) OR ABOUT 25 SAUSAGES

5 kg (11 lb) fresh boneless, skinless tuna trim
2 kg (4 lb 6 oz) cold-smoked salmon fillet, skinless
500 g (1 lb 2 oz) Cured Fish Fat (page 250)
100 g (3½ oz) table salt
17 g (⅔ oz) cure #2
10 g (¼ oz) onion powder
20 g (¾ oz) grated garlic
10 g (¼ oz) sodium erythorbate
400 g (14 oz/2⅔ cups) peeled raw pistachio nuts
ox casings

Place the tuna, smoked salmon and fish fat in the freezer. Once below 0°C (30°F), mince through a 13 mm (½ in) plate on a commercial meat grinder.

Wearing gloves, mix the ground fish with all the other ingredients except the pistachios until sticky. Add the pistachios and give a quick mix to distribute evenly. Place the sausage mix in the refrigerator for at least 1 hour until ready to fill the casings.

Cover and soak the casings in cold water for 1 hour prior to filling.

Set up a sausage filler. Temperature check the fish during mincing and filling – you are aiming to stay below 8°C (45°F). While filling the casings, ensure the work surface is damp so the sausages don't stick. This can be done with a spray bottle of water to mist over the bench.

Using butcher's twine, tie the sausages at about 15 cm (6 in) lengths, making sure they are filled tightly to avoid separation.

Poke the sausages all over with a needle to allow moisture to escape; you want a weight loss of about 30 per cent.

Hang sausages on butcher's hooks and store in a well-ventilated fridge to dry. (These sausages are not fermented, only cured, so a salami cabinet is not necessary.) Allow 4 weeks before slicing open to try – depending on the humidity of your fridge, this process may take less or more time.

COLD-SMOKED FISH HAM

This cure translates well between a number of different species. My suggestion, however, is to work with a fatty fish in peak condition to extend the shelf life of a beautiful fish in the form of a cold cut or ham.

If you don't want to commit a whole fillet to this recipe, simply cure the belly of the fish. The fish shown here is a striped marlin, and one of the loins has been removed, cured, smoked, dried for 2 weeks and then thinly sliced on a meat slicer.

MAKES 1 X 5 KG (11 LB) HAM

4.75 litres (167 fl oz/19 cups) water
250 g (9 oz) table salt
25 g (1 oz) cure #1
5 kg (11 lb) boneless kingfish, tuna, swordfish or marlin, skinned and sinew removed

Stir together the water, salt and cure #1 until completely dissolved.

Place the trimmed fish in brine and refrigerate for at least 12 hours but no more than 24 hours.

Remove from the brine and dry well using paper towel. Prepare a smoker and cold smoke for 2 hours. Remove from the smoker and chill in the refrigerator.

Once cold, this ham can be either kept fresh like this and thinly sliced and served or be cooked. One possibility would be to cook the fish to an internal temperature of approximately 45°C (115°F) in a low-temperature oven and then slice thickly and drape over hot toast with raw onion, pickles and fresh herbs.

MARLIN 'NDUJA

Ensure your marlin is incredibly fresh when making a recipe like this – see it as a way to utilise accumulated trim, not something that you would purchase a kilogram of perfect centre-cut marlin for. Like the salamis, sausages and other products featured here, this is an attempt to clean up the scraps and transform them into something delicious.

MAKES 1.4 KG (3 LB 1 OZ)

1 kg (2 lb 3 oz) boneless, skinless marlin trim
300 g (10½ oz) rendered fish fat
3 garlic cloves, minced
100 g (3½ oz) 'Nduja spice mix (see below)
3 g (⅛ oz) cure #1

'NDUJA SPICE MIX

550 g (1 lb 3 oz) smoked paprika
100 g (3½ oz) sweet paprika
170 g (6 oz) table salt
20 g (¾ oz) ground nutmeg
100 g (3½ oz) chilli flakes
30 g (1 oz) ground coriander seeds
20 g (¾ oz) ground cumin seeds
20 g (¾ oz) ground black pepper
30 g (1 oz) Vegeta stock powder
30 g (1 oz) ground sumac

To make the 'nduja spice mix, combine all the ingredients in a small bowl; the excess can be stored in a clean plastic container for later use.

Pass the trimmed marlin through a meat grinder on a coarse setting. Transfer to a large bowl, add the rendered fish fat, minced garlic, spice mix, cure #1 and a pinch of salt. Using gloves, mix the 'nduja to combine (this will take about 5 minutes to gain a slightly sticky, firm result).

Using a sausage filler, the 'nduja can be stored in collagen casings that are tied in butcher's twine and then hung in a coolroom to increase flavour and reduce some of the moisture. Alternatively, it can be stored in a stainless steel container or cryovac bag in the refrigerator to be consumed sooner. The 'nduja can be fried in a pan and used in place of a traditional meat-based 'nduja or alternatively eaten as a fresh product.

TUNA SALAMI PICCANTE

This is a fiery version of our original tuna salami that utilises the trim from smoked salmon as a way of bringing smokiness to the final product without the need to smoke the finished salami.

MAKES 4.5 KG (9LB 15 OZ) OR ABOUT 25 SAUSAGES

5 kg (11 lb) boneless, skinless tuna trim
2 kg (4 lb 6 oz) cold-smoked salmon
500 g (1 lb 2 oz) Cured Fish Fat (page 250)
100 g (3½ oz) table salt
17 g (½ oz) cure #2
15 g (½ oz) onion powder
15 g (½ oz) garlic powder
22 g (¾ oz) kibbled black pepper
40 g (1½ oz) smoked paprika
7 g (⅛ oz) ground coriander seeds
15 g (½ oz) ground fennel seeds
45 g (1½ oz) ground chilli
10 g (¼ oz) sodium erythorbate
ox casings

Place the tuna, smoked salmon and fish fat in the freezer. Once below 0°C, mince through a 13 mm (½ in) plate on a commercial meat grinder.

Wearing gloves, mix the mince with all the other filling ingredients until sticky, about 10 minutes. Place the sausage mix in the refrigerator for at least 1 hour until ready to fill the casings.

Soak the casings in water for 1 hour prior to filling.

Set up a sausage filler. Temperature check the fish mixture during mincing and filling, aiming to stay below 8°C (45°C). When filling the casings, make sure the work surface is damp so the sausages don't stick. This can be done with a spray bottle of water to mist over the bench.

Using butcher's twine, tie the sausages at about 15 cm (6 in) lengths, making sure they are filled tightly to avoid separation. Poke the sausages all over with a needle to allow moisture to escape; you want a weight loss of about 30 per cent.

Hang the sausages on butcher's hooks in a well-ventilated fridge to dry. (These sausages are not fermented, only cured, so a salami cabinet is not necessary.) Allow 4 weeks and then slice open to try. Depending on the humidity of your fridge, this process may take less or more time.

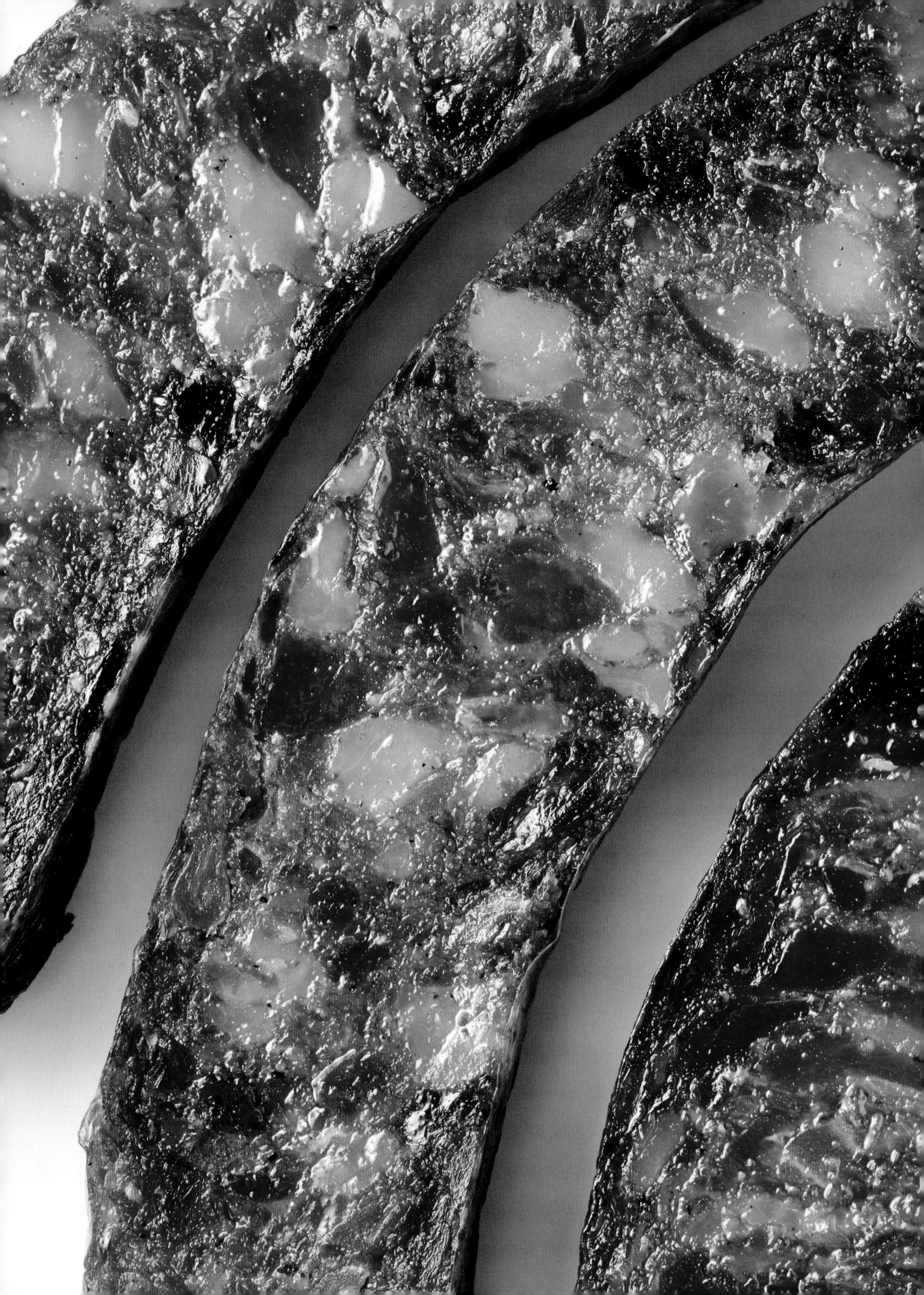

TUNA IN THE STYLE OF SAUCISSON SEC

Although we cannot say that this is a saucisson sec, we can definitely reference it here as our starting point and inspiration for this recipe. By increasing the percentage of fat and also introducing a fatty belly section of the tuna, this salami is wonderfully luxurious and worth waiting the additional week for it to be ready.

MAKES 4.5 KG (9 LB 15 OZ) OR ABOUT 25 SAUSAGES

- 5 kg (11 lb) boneless, skinless tuna trim
- 2 kg (4 lb 6 oz) cold-smoked tuna belly (see Cold-smoked Fish Ham, page 162)
- 500 g (1 lb 2 oz) Cured Fish Fat (page 250)
- 120 g (4½ oz) table salt
- 20 g (¾ oz) cure #2
- 20 g (¾ oz) ground black pepper
- 225 g (8 oz) minced garlic
- 10 g (¼ oz) ground nutmeg
- 10 g (¼ oz) ground cumin seeds
- 10 g (¼ oz) sodium erythorbate
- ox casings

Place the tuna, smoked tuna belly and fish fat in the freezer. Once below 0°C (30°F), mince through a 13 mm (½ in) plate. Mix together with all the other ingredients. Place the sausage mixture in the refrigerator for at least 1 hour until ready to fill casings.

Soak the casings in water for 1 hour prior to filling

Set up a sausage filler. Temperature check the mixture during mincing and filling, aiming to stay below 8°C (45°F). When filling the casings, make sure the work surface is damp so the sausages don't stick. This can be done with a spray bottle of water to mist over the bench.

Using cotton butcher's twine, tie the sausages at intervals about 15 cm (6 in) long, making sure they are filled tight to avoid separation. Poke the sausages all over with a needle to allow moisture to escape; you want a weight loss of about 30 per cent.

Hang the sausages on butcher's hooks and store in a well-ventilated fridge to dry. (These sausages are not fermented, only cured, so a salami cabinet is not necessary.) Allow 5 weeks and then slice open to try – depending on the humidity of your fridge, this process may take less or more time. Once the desired texture is achieved, this salami can be cryovacked and stored in the fridge to remove the risk of further moisture loss.

These salami are best enjoyed cut slightly thick and served with good pickles or olives.

TUNA BRESAOLA

Unlike other recipes utilising the trim or sundries from a single fish, this intentionally carves out the centre of a loin of tuna to capture its perfection and texturally adjust it into something very similar to a beef bresaola. Be sure to have a plan for the trim before you set out to produce this recipe.

Ensure that you only slice this product when you wish to consume it as it will oxidise quickly.

MAKES 1.5 KG (3 LB 5 OZ)

1 x 4 kg (8 lb 13 oz) trimmed A-grade yellowfin tuna loin
fresh bay leaves
whole cinnamon sticks
collagen sheeting

CURE

300 g (10½ oz) ground black pepper
150 g (5½ oz) ground fennel seeds
150 g (5½ oz) table salt
100 g (3½ oz) caster (superfine) sugar
100 g (3½ oz) cure #2

In a large mixing bowl, combine all the ingredients for the cure.

Take the tuna loin and, using a sharp chef's knife, round off the ends and circumference of the loin into an evenly sized barrel weighing approximately 2.5 kg (5½ lb). (Any trim from this process should be used for another application, eg Marlin 'Nduja (see page 165), Tuna Bolognese (see page 241), Tuna Patties (see page 222) or the salami recipes.)

Before adding the cure to the tuna, weigh the loin to calculate the ratio correctly. Using a small coarse sieve, dust the barrel of tuna with 40 g (1½ oz) of cure per kilogram (2 lb 3 oz) of tuna along with two bay leaves and half a cinnamon stick, making sure the cure is evenly distributed. Leave to cure in a sterilised storage container in the refrigerator for 4 days, then turn and leave for a further 4 days.

On day 8, remove the cinnamon and bay leaves. Roll the log in collagen sheeting, allowing a good amount of overlap, then truss each log using butcher's twine. (Use a looping method like trussing a porchetta, then lock in the other side, over and under on each loop. When you reach the top, tie off with a loop to hang the bresaola.) Hang on a butcher's hook in the fridge for 3–4 weeks, depending on its thickness.

The finished result will be firm to the touch, with a defined cured edge right the way around and a desirable waxy centre. This tuna bresaola is the type of ham that you would slice straight onto the serving plate using a meat slicer and then serve immediately. Once sliced, the remaining bresaola can be cryovacked to remove the risk of it oxidising and losing its beautiful colour.

FISH BACON

Fish bacon might sound as enticing as plant-based bacon to most, but this creation has been the catalyst for some of our most delicious menu items at Saint Peter, Fish Butchery and Charcoal Fish.

One dish that this bacon specifically lands on is our yellowfin tuna cheeseburger. Crowned with a halo of the crispiest, smokiest bacon that to most is nearly unrecognisable as fish, it often shocks pescatarians wondering if they've mistakenly ordered pork. The most suitable fish is one that carries a rich amount of intramuscular fat, and this cure works best with boneless fillets as opposed to whole fish on the bone.

MAKES 2.18 KG (4 LB 13 OZ) OF CURE

boneless, skinless fish fillet (kingfish, cod, swordfish, tuna, sea mullet)

DRY BACON CURE
(use 130 g (4½ oz) per kilogram (2 lb 3 oz) of fish)

500 g (1 lb 2 oz) caster (superfine) sugar
1 kg (2 lb 3 oz) table salt
30 g (1 oz) ground star anise
150 g (5½ oz) ground cumin seeds
50 g (1¾ oz) cure #1
150 g (5½ oz) ground black pepper
150 g (5½ oz) ground coriander seeds
150 g (5½ oz) ground fennel seeds

Weigh the fish, sprinkle evenly with the appropriate ratio of cure mix and place on a stainless steel tray. Allow the fish to cure for 7 days in the refrigerator, turning each day, then rub off the excess cure.

In a smoker, cold smoke the fish for 4 hours.

Hang the fillets on butcher's hooks in a fan-forced refrigerator for between 7 and 14 days, depending on the humidity of your fridge. Once the fillet has reduced by 30 per cent of its total raw weight, the bacon is ready. Set up a sharp meat slicer and cut the finished fish bacon into thin slices that best suit the application. Alternatively, the bacon can be cut with a knife for a slightly thicker finish. Once sliced, any remaining bacon can be returned to the refrigerator and either hung or cryovaced to avoid too much more moisture being lost.

SUJUK

This is our interpretation of sujuk sausage (which is traditionally made with beef or lamb and fermented) utilising the trim of tuna. We use tuna because it is lean and, unlike other sausages and salamis in this book, this one contains no fat at all. Once cooked, this sujuk is delicious thickly sliced and grilled over hot charcoal.

MAKES 2.5 KG (5½ LB) OR 3 SAUSAGES

- 2.5 kg (5½ lb) boneless, skinless tuna trim
- 70 g (2½ oz) table salt
- 4 g (⅛ oz) cure #1
- 8 g (¼ oz) ground black pepper
- 25 g (1 oz) ground cumin
- 5 g (⅛ oz) ground allspice
- 10 g (¼ oz) roasted red pepper puree
- 25 g (1 oz) garlic, peeled and minced
- 3 x 55 mm (2¼ in) plastic casings

Preheat a combination steam and convection oven to a 65°C (150°F) steam.

Place the tuna in the freezer and, once below 0°C (30°F), mince through a 2 mm (⅛ in) plate. Combine the ground tuna with all of the remaining ingredients in a large mixing bowl. Place in the refrigerator until ready to fill.

Place the sujuk mix in a sausage filler and fill the plastic casings. Tie the casings off very tightly with butcher's twine and hang on butcher's hooks in the combination oven. Steam for approximately 20 minutes, or until the internal temperature reaches 55°C (130°F).

Remove from the oven. Allow to cool for 1 hour before refrigerating, ensuring the sausages remain hung while cooling to maintain their cylindrical shape.

Once cooled, the sujuk can be sliced and eaten as is or, alternatively, fried.

BASIC SAUSAGE

This basic sausage recipe is a fantastic starting place to achieve just about any sausage you can imagine. With this recipe in your tool belt, you will be able to eliminate a huge amount of fish waste. I genuinely hope that if just one recipe is taken from this book it will be this.

**MAKES 2.5 KG (5½ LB)
OR ABOUT 22 SAUSAGES**

1 kg (2 lb 3 oz) boneless, skinless salmon trim
1 kg (2 lb 3 oz) boneless, skinless white fish trim (preferably cod, flathead, gurnard, snapper, non-scombroid species)
400 g (14 oz) Cured Fish Fat (page 250)
150 g (5½ oz) quinoa flakes
25 g (1 oz) table salt
15 g (½ oz) ground fennel seeds
15 g (½ oz) ground black pepper
6 g (⅛ oz) cure #1
1 hog casing

Place the salmon, white fish and fish fat in the freezer. Once below 0°C (30°F), mince through a commercial meat grinder with a 10 mm (½ in) plate.

Wearing gloves, mix the mince with all the other filling ingredients until sticky, about 10 minutes. Place the sausage mix in the refrigerator for at least 1 hour until ready to fill the casing.

Soak the casing in water for 1 hour prior to filling.

Set up a sausage filler. Temperature check the fish mixture during mincing and filling, aiming to stay below 8°C (45°F). When filling the casing, make sure the work surface is damp so the sausages don't stick. This can be done with a spray bottle of water to mist over the bench.

Poke the sausages with a needle if you see any obvious air pockets, then link the sausages into approximately 12 cm (4¾ in)lengths by twisting the filled casing by hand.

Hang sausages on butcher's hooks in a refrigerator. If keeping longer than 2 days, cover the sausages to prevent overdrying.

These fresh sausages can be simply pan-fried in a little oil or, alternatively, grilled over coals until just done. Subjecting the sausages to too high a heat can cause the skins to break. If cooked for too long, you run the risk of drying out the sausage as there is naturally less fat in this sausage than a conventional meat-based variety.

CHIPOLATAS

Traditionally made with coarsely ground pork, this interpretation of a classic chipolata uses the fattiness of salmon and cured fish fat to give the sausage moisture and snap. While the seasonings listed below are one suggestion, see it as a blank canvas that can be interpreted in a number of different ways once you have mastered the technique of sausage making.

MAKES 5.4 KG (11 LB 14 OZ) OR ABOUT 88 SAUSAGES

2 kg (4 lb 6 oz) boneless, skinless salmon trim
2 kg (4 lb 6 oz) boneless, skinless white fish trim
800 g (1 lb 12 oz) Cured Fish Fat (page 250)
300 g (10½ oz) quinoa flakes
40 g (1½ oz) table salt
100 g (3½ oz) smoked paprika
10 g (¼ oz) sweet paprika
3 g (⅛ oz) sumac
3 g (⅛ oz) Vegeta stock powder
3 g (⅛ oz) ground coriander seeds
3 g (⅛ oz) ground cumin seeds
2 g (⅛ oz) ground black pepper
2 g (⅛ oz) grated fresh nutmeg
12 g (¼ oz) cure #1
2 lamb casings

Place the salmon, white fish and fish fat in the freezer. Once below 0°C (30°F), mince in a commercial meat grinder with a 10 mm (½ in) plate. Mix together with all the other filling ingredients until sticky, approximately 10 minutes. Place the sausage mix in the refrigerator for at least 1 hour until ready to fill casings.

Soak the casings in water for 1 hour prior to filling.

Set up a sausage filler. Temperature check the mixture during mincing and filling, aiming to stay below 8°C (45°F). When filling the casings, make sure the work surface is damp so the sausages don't stick. This can be done with a spray bottle of water to mist over the bench.

Link the sausages into approximately 8 cm (3¼ in) lengths by twisting the filled casings by hand.

Hang the sausages on butcher's hooks in a refrigerator. If keeping longer than 2 days, cover the sausages to prevent overdrying.

These fresh chipolatas can be simply pan-fried in a little oil or, alternatively, grilled over coals until just done. Subjecting the sausages to too high a heat can cause the skins to break. If cooked for too long, you run the risk of drying out the sausage as there is naturally less fat in this sausage than a conventional meat-based variety.

COLD-SMOKED TUNA TAIL HAM ON THE BONE

This preparation came about in response to a lack of creativity when using a fish tail more intentionally. Too often the tail piece of fillet is considered less desirable, so before it comes off the bone, put it to work as a larger piece on the bone. Curing and smoking this section of the fish increases its shelf life drastically, giving you more time and greater opportunity to use the whole fish. This is a versatile recipe that is better applied to larger fish. Select fish with a generous amount of intramuscular fat for best results.

MAKES 500 G (1 LB 2 OZ) OF CURE

tuna tail
collagen casing

HAM CURE

75 garlic cloves, bruised, skin on
1 rosemary sprig
2 fresh bay leaves
15 g (½ oz) whole fennel seeds
15 g (½ oz) whole black peppercorns
300 g (10½ oz) table salt
100 g (3½ oz) caster (superfine) sugar
50 g (1¾ oz) cure #1

To prepare this cut, lay a whole tuna across a clean work surface and, using a sharp knife, cut just behind the anal fin and all the way through. Remove the caudal fin completely and trim off 5 cm (2 in) of flesh to expose the bone. Once the tail bone is exposed, clean away the flesh from it for aesthetic purposes. Using a sharp knife, remove the skin from the tail, ensuring that minimal flesh is left on it.

Combine the ham cure ingredients in a stainless steel bowl; any excess can be stored in a clean plastic container for later use. Rub the skinless tail of the tuna with the cure mix at the ratio of 110 g (4 oz) of cure per kilogram (2 lb 3 oz) of tuna. Cure the tuna tail uncovered in a sterilised storage tub in the refrigerator for 8–10 days, turning over every 2 days.

After this time has passed, wearing gloves, rub the excess cure from the tail.

In a smoker, cold smoke the tail for 2 hours, then remove and refrigerate until thoroughly chilled.

Wrap the tail in collagen casing. (This is a preventative measure to avoid case hardening and minimise excess moisture loss during drying.) Tie the collagen to the bone with butcher's twine, running the twine under the tuna and around the bone three times with equal spacing creating six lines of twine. On the fourth run, stop at the bottom and cut the twine, leaving 2 metres (6½ ft) of extra twine. Start weaving the twine in and out to create a spider web effect until you reach the bone at the top, then tie off with a loop for the hook. Using a pin, puncture the collagen casing surrounding the tail for ventilation.

Hang to dry in a conventional fan-forced coolroom set to 2°C (35°F) for 4–8 weeks. The final texture should resemble that of a good-quality smoked leg ham and it can be consumed as is, cooked cut into lardons and pan-fried or thinly sliced and used for sandwiches. After all the meat has been cut from it, the smoked tuna tailbone can be used in stock to create a pea and tuna tail ham soup.

FRANKFURTS

Again, we're not reinventing the wheel with this, merely tapping into a method that is critical in the world of meat butchery. While all of us have likely had a hot dog, the challenge was to see if a fish hot dog could be as delicious. The use of prawns (shrimp), scallops and the trim of salmon turn the old-fashioned dog into something far more memorable.

MAKES 4.4 KG (9 LB 11 OZ) OR ABOUT 40 SAUSAGES

- 2 kg (4 lb 6 oz) roe on scallop meat, cleaned
- 1 kg (2 lb 3 oz) raw, shelled and cleaned prawn (shrimp) meat
- 800 g (1 lb 12 oz) skinless, boneless salmon belly
- 500 g (1 lb 2 oz) rendered fish fat
- 60 g (2 oz) table salt
- 4 egg whites
- red collagen casings

Combine the scallop, prawn, fish meat and rendered fish fat and add the salt. Divide this mixture into four batches and blend the first in a Robot-Coupe processor until almost smooth. Add in one of the egg whites and continue to blend until the mixture is smooth and mousse-like. Remove to a large mixing bowl and proceed to blend the remaining batches, adding one egg white to each batch.

Use a sausage filler to fill and tie the mixture in red collagen casings. Refrigerate for at least 2 hours prior to cooking.

Place the frankfurts in a pot of cold water over a low heat and slowly bring up until the water is 60°C (140°C). Turn off the heat and let the frankfurts sit in the hot water for 5 minutes.

Drain the sausages from the cooking water and chill on trays covered with a damp clean cloth in the refrigerator.

To heat the frankfurts, bring a large pot of water to the boil and remove from the stove. Place the frankfurts in the pot and place the lid on top. Stand for approximately 10 minutes until thoroughly heated through. Remove carefully and serve on a warm bun with your favourite condiments.

PATE DE CAMPAGNE

This coarsely cut ensemble of fish and offal is a remarkable way of creating a product out of potentially discarded fish that is recognisable to those of us who appreciate the art and craft of meat-based charcuterie. It is another great recipe to experiment with different flavours and seasonings to make it your own.

MAKES 6 KG (13 LB 4 OZ)

- 5 kg (11 lb) skinless, boneless white fish trim (cod, gurnard, dory)
- 150 g (5½ oz) coarsely diced Fish Bacon (page 174), plus a further 500 g (1 lb 2 oz) sliced on its longest side
- 500 g (1 lb 2 oz) Cured Fish Fat (page 250), coarsely diced
- 200 g (7 oz) best fish liver, trimmed and coarsely diced
- 600 g (1 lb 5 oz) brown onions, finely diced
- 50 g (1¾ oz) garlic cloves, grated
- 20 g (¾ oz) thyme leaves, finely chopped
- 20 g (¾ oz) rosemary leaves, finely chopped
- 20 g (¾ oz) ground white pepper
- 12 g (¼ oz) cure #1
- 50 ml (1¾ fl oz) cooking brandy
- sea salt flakes, to taste

Preheat a combination steam and convection oven to 90°C (195°F).

Place the white fish in the freezer. Once below 0°C (30°F), mince through a 13 mm (½ in) plate. Combine the chilled ground white fish in a large mixing bowl with all of the remaining ingredients except the long fish bacon slices. Season with sea salt to taste.

Line a 33 x 8 x 8 cm (13 x 3¼ x 3¼ in) terrine mould with the pre-sliced fish bacon, ensuring that enough bacon hangs over the sides of the terrine to later fold over and cover the top. The bacon also needs to be overlapping in the terrine to ensure the filling is completely encased.

Press the mixture firmly into the mould and fold the overhanging bacon across the top.

Place a square of baking paper over the bacon and then a square of aluminium foil to cover the terrine well, crimping the foil to seal.

Place the terrine mould on a baking tray and put in the oven, cooking until the internal temperature reaches approximately 55°C (130°F). Remove from the oven when cooked, take off the foil and allow to cool for approximately 15 minutes.

Place a rectangle of cardboard that matches the terrine width and length and set it on top of the baking paper. Press overnight using an approximately 2 kg (4 lb 6 oz) weight.

The next day, the terrine can be turned out onto a cutting board and sliced to serve.

BRAWN

This recipe specifies the weight of picked and cooked head meat that you need, so the number of heads required to achieve this will be subject to their size and yield. To cook the fish heads, set a steam oven to 70°C (160°F) and arrange the heads in one even layer. Ensure that the heads are not overcooked as they will lose the juices and setting qualities that are essential to a beautifully dense and gelatinous finish here.

MAKES 1

- 1.5 kg (3 lb 5 oz) cooked and picked head meat and their juices (preferably a gelatine- and fat-rich fish like cod, monkfish or coral trout)
- 150 g (5½ oz) French shallots, peeled and finely diced
- 170 g (6 oz) salted tiny capers, rinsed
- 85 g (3 oz) Dijon mustard
- zest of 2 lemons
- 3 tablespoons finely chopped tarragon leaves
- salt and cracked black pepper, to taste

Combine the still-warm cooked head meat with the other ingredients, adding the herbs and seasoning to taste.

Line a 33 x 8 x 8 cm (13 x 3¼ x 3¼ in) terrine mould with a double layer of plastic wrap, ensuring there is enough overhang to wrap back over the top of the terrine. Press the mixture firmly into the mould.

Wrap the plastic over the long sides of the terrine to keep the shape but leave it open at the shorter ends to allow any excess liquid to escape while pressing.

Place a thick piece of cardboard cut to size over the top of the terrine and weigh it down evenly with something heavy. Allow it to set in the refrigerator overnight before using.

Use a sharp kitchen knife to slice the terrine and keep covered when refrigerated to avoid the terrine drying out.

RILLETTE

Like a traditional pork or duck rillette, this is more a mechanism for fully utilising even the most insignificant of trim. In this recipe I have suggested using our fish ham as a starting point, as the cure gives the fish a fantastic flavour. This rillette is rich and should be enjoyed with plenty of pickles and fresh bread.

MAKES 1.5 KG (3 LB 5 OZ)

1.5 kg (3 lb 5 oz) ghee
12 juniper berries, crushed
1 head of garlic, halved
6 thyme sprigs
4 rosemary sprigs
2 fresh bay leaves
1.5 kg (3 lb 5 oz) Cold-smoked Fish Ham (page 162)
6 French shallots, finely diced
salt and freshly cracked black pepper, to taste
150 ml (5 fl oz) brandy
100 ml (3½ fl oz) rendered fish fat (if using, otherwise replace with additional ghee)

Start by heating the ghee, juniper berries, garlic and herbs up to 40°C (105°F) in a wide-based saucepan over a medium heat.

Place a wire rack in the base of the pan (to avoid the smoked fish getting too hot on the base) and reduce the heat to low, holding the temperature between 40–45°C (105–115°F).

Add the ham to the saucepan and poach gently until the flesh comes apart easily, approximately 15 minutes depending on the thickness of the fish. Once cooked, remove it from the ghee and set aside. Cool the ghee down slightly and then return the fish to the pan to continue to rest and gain more flavour.

Once cooled, remove the fish and strain well. Using two forks, break up the fish into a consistency that is fine but not mushy.

Strain the ghee of its aromatics and measure 50 ml (1¾ fl oz) into a small saucepan. Add the diced shallots to the ghee and cook over a low heat until tender and translucent. Add the shallots and ghee to the shredded fish. Season to taste and add the brandy to the mixture.

Pot this mixture into ramekins or a terrine of choice and press down firmly using a spoon or spatula to ensure there are no gaps. Strain another 100 ml (3½ fl oz) of the aromatic cooking ghee into a measuring jug and add 100 ml (3½ fl oz) of rendered fish fat (if using). Pour a thin layer of this fat over the ramekins of rillette. Store in the refrigerator overnight to set. The rillettes are best served with grilled sourdough and pickles.

OPINEL
INOX

LIVER PÂTÉ

I absolutely love liver pâté and believe that if the fish livers are in excellent condition and the pâté is well made, this will be a recipe you return to over and over. Be sure to keep the livers nice and pink as if they cook too far, the pâté can end up too grainy and the finished colour can be more grey than pink.

MAKES 1 KG (2 LB 3 OZ)

1 kg (2 lb 3 oz) fish livers, trimmed and patted dry
2 g (⅛ oz) cure #1
500 g (1 lb 2 oz) diced butter, at room temperature
4 French shallots, peeled and sliced
200 ml (7 fl oz) port
4 thyme sprigs
6 fresh bay leaves
160 g (5½ oz) ghee
salt and pepper, to taste

Season the livers with some salt (indicative of how much seasoning you want the pâté to have) and the cure #1 and allow to sit on a tray in the refrigerator uncovered for 10 minutes prior to cooking.

In a frying pan set over a medium heat, add 75 g (2¾ oz) of the butter and cook the shallots until slightly caramelised, approximately 15 minutes.

Add in the port, thyme and bay leaves, being careful as the port will flame. Boil and reduce over a medium heat until syrupy, approximately 10 minutes. Allow these glazed ingredients to cool while you start to cook the livers.

Set a heavy-based cast-iron pan over a high heat. Add 40 g (1½ oz) of the ghee to the pan and allow it to reach a light haze, then carefully add approximately 250 g (9 oz) of the livers. The ambition here is to brown the livers on both sides but keep a very rare pink interior – it should be a matter of a few seconds on each side. The flavour created when caramelising the livers will be critical to the final outcome. After each batch is browned off, set the livers aside in a colander to drain. Wipe the pan out and repeat until all the livers are cooked.

Remove the hard herbs from the reduction of port and shallots. Combine the cooked livers with the cooled reduction and add further salt and pepper.

Using a jug blender, blend the liver mixture in four batches until completely smooth. While still blending, drop a quarter of the remaining butter, piece by piece, into each batch. Tip each finished batch into a fine sieve set over a bowl, then continue blending the remaining batches.

Pass the mixture through the fine sieve with a ladle to help force it through, season again to taste and then pour into your chosen ramekins, containers or moulds.

A thin layer of jelly or clarified butter can be poured over the top of this pâté to avoid oxidisation.

TUNA, PISTACHIO AND FIG TERRINE

Fruit and fish? While the combination is as frowned upon as cheese and fish, I implore you to reconsider. The tuna in this recipe could easily be mistaken for duck or pork, and it was screaming out for natural sweetness. So the inspiration came from the use of dried fruits in meat-based terrines, which offer both subtle sweetness but also relief when eating something quite fatty. In this recipe, the lateral swimming muscle of the tuna is used to impart a more savoury characteristic along with a darker, meatier colour. Most will identify this part of a fish as the 'bloodline', which is not actually the case.

Enjoy this one with grilled sourdough and a big spoonful of mustard.

MAKES 6 KG (13 LB 4 OZ)

300 g (10½ oz) tuna lateral swimming muscle
1 tablespoon table salt, for curing
5 kg (11 lb) boneless, skinless tuna trim
500 g (1 lb 2 oz) Cured Fish Fat page 250), diced
600 g (1 lb 5 oz) brown onions, finely diced
50 g (1¾ oz) garlic cloves, grated
250 g (9 oz) pistachio nuts, peeled
100 g (3½ oz) dried figs, sliced
20 g (¾ oz) thyme leaves, finely chopped
20 g (¾ oz) rosemary leaves, finely chopped
20 g (¾ oz) ground white pepper
12 g (¼ oz) cure #1
50 ml (1¾ fl oz) brandy
salt and cracked black pepper, to taste

Preheat a combination steam and convection oven to 90°C (195°F).

Salt the lateral swimming muscle of the tuna in the tablespoon of salt for approximately 30 minutes, then finely dice.

Place the tuna trim in the freezer. Once below 0°C (30°F), mince through a 13 mm (½ in) plate. Combine the ground tuna and diced lateral swimming muscle with all of the remaining ingredients in a large mixing bowl. Season with sea salt to taste.

Line a 33 x 8 x 8 cm (13 x 3¼ x 3¼ in) terrine mould with baking paper, ensuring that enough paper hangs over the sides of the terrine to later fold over and cover the top. Press the mixture firmly into the mould and fold the overhanging paper across the top. Place a square of aluminium foil to cover the terrine well, crimping the foil to seal. Place the terrine mould onto a baking tray and place in the oven, cooking until the internal temperature reaches approximately 55°C (130°F). Remove from the oven when cooked, take off the foil and allow to cool for approximately 15 minutes.

Place a rectangle of cardboard that matches the terrine width and length and set on top of the baking paper. Press overnight using an approximately 2 kg (4 lb 6 oz) weight.

The next day, the terrine can be turned out onto a cutting board and sliced to serve.

PASTRIES, PATTIES AND CRUMBED GOODS

TUNA MINCEMEAT PIE

While there are another two white sauce-based pie recipes in this book, I wanted to include this 'meat pie' that interprets tuna as beef and suspends the ground tuna in a dark fish gravy that any tradie would be happy to squeeze tomato sauce onto for their lunch.

MAKES 75 PIES

- 3.5 kg (7 lb 12 oz) boneless, skinless tuna trim
- 125 g (4½ oz) salted butter
- 3.5 kg (7 lb 12 oz) brown onions, diced
- 400 g (14 oz) garlic cloves, grated
- 50 g (1¾ oz) fresh thyme, finely chopped
- 8 g (¼ oz) crushed black peppercorns
- 8 g (¼ oz) crushed white peppercorns
- 8 g (¼ oz) ground nutmeg
- 3 g (⅛ oz) ground star anise
- 3.5 litres (118 fl oz/14 cups) Fish Gravy (page 249)
- 75 g (2¾ oz) plain (all-purpose) flour
- 1 kg (2 lb 3 oz) Sour cream pastry (page 212), rolled to 4 mm (¼ in) thickness and chilled
- 4 egg yolks, beaten, for egg wash

Place the tuna in the freezer. Once below 0°C (30°F), mince through a 13 mm (½ in) plate. Set aside in the refrigerator until required.

In a large pan set over a medium heat, add the butter and allow it to start to foam without browning. Add the onions and cook until tender, golden and caramelised, approximately 30 minutes. Add the garlic, thyme, peppercorns and spices and cook for a further 15 minutes until fragrant. Next add the fish gravy.

In a large mixing bowl, season the tuna with the plain flour.

Taking small amounts at a time to avoid lumps forming, whisk the floured ground tuna into the sauce. Once all the tuna is added and well incorporated, season well and cook for a further 10 minutes so the stock can thicken. The mix shouldn't be too loose – there needs to be enough sauce to bind the meat and keep the pie juicy but not runny. Refrigerate for 24 hours to cool and set. (If your mix hasn't set for long enough you will run into trouble when crimping the pies.)

On a clean and cold work surface, lay out the pastry to cut it to size. You will be lining the base and top of these pies, so cut a larger circle to cover the base and come up the sides with 1 cm (½ in) of overhang, and a smaller sheet to cover the top of the pie. For this we use circular 142 ml (4½ fl oz) dishes measuring 9 cm (3½ in) in diameter and 2.1 cm (¾ in) deep.

Once the pastry is cut to size, line the bases of the pie dishes with the larger rounds of pastry. Gently push the pastry into the edges so there is as much space as possible for the filling. Spoon 100 g (3½ oz) of the chilled filling into each pie and spread it out to fill the gaps. Place the smaller round of pastry directly over the top of the filling and crimp the overhanging pastry from the base layer over the top to create a neat seal.

With a pastry brush, paint the tops with the egg yolks, ensuring an even layer.

Place in the refrigerator for 30 minutes to dry the egg glaze. Glaze the pie again before baking to create a bronzed finish. Bake in a 185°C (365°F) oven for 25–30 minutes.

INDIVIDUAL FISH PIE

This recipe is an excellent way to use up the trim from any mild-flavoured flaky white fish fillets. When we portion fish fillets for retail, we often square off a little towards the tail, the belly and the top of the fillet. You can collect and skin these trimmed pieces and they can be combined and used in this pie. It's also a great vehicle to carry the offal of fish as well.

MAKES 25 PIES

1.5 kg (3 lb 5 oz) skinless, boneless white fish trim (any white, flaky and mild variety works well)
8 g (¼ oz) salt
2 litres (68 fl oz/8 cups) brown fish stock
200 g (7 oz) unsalted butter, diced into 2 cm (¾ in) cubes
200 g (7 oz/1⅓ cups) plain (all-purpose) flour
pinch of freshly grated nutmeg
250 g (9 oz) fresh corn cobs
250 g (9 oz) baby spinach leaves, picked
1 kg (2 lb 3 oz) Sour cream pastry (page 212), rolled to 4 mm (¼ in) thickness and chilled
4 egg yolks, beaten, for egg wash
salt and cracked black pepper, to taste

Chop the fish trim into 1.5 cm (½ in) bite-sized pieces and place on a deep tray in a single layer. Evenly sprinkle with the 8 g (¼ oz) of salt and toss to combine well. Place the tray of seasoned fish in the refrigerator overnight to firm up.

The next day, preheat a steam oven to 70°C (160°F). Remove the lightly salted fish from the fridge and, once the steam oven is ready, place into the oven for 4–7 minutes until just cooked through. Remove from the oven and place in the refrigerator to cool completely in its juices.

Place the fish stock into a pot over a low heat until it just comes to a boil. Take care not to let the stock boil for longer as it will begin to reduce and the filling will end up too thick. For this step, we are just aiming to get the stock at its hottest point.

Meanwhile, place the butter in a saucepan and set over a medium heat. Once the butter has fully melted, add the flour. Combine the two together into a roux and stir for 2–3 minutes, or until the rawness of the flour is cooked out. Remove from the heat.

As soon as the stock comes to the boil and the roux is ready to go, take the pot of stock off the heat momentarily and gently spoon the roux in. Use a whisk to break up the roux and combine well into the liquid. Once the lumps of roux have been fully incorporated into the liquid, place the pot back over a medium-low heat and continue to whisk until the sauce begins to bubble. Continue whisking to avoid scorching the bottom while you allow it to cook out for a further 5 minutes. Mix in the nutmeg and season with salt and cracked black pepper to your taste. Transfer your finished velouté into a deep container or tray, cover with plastic wrap and place in the refrigerator to cool.

Bring a pot of salted water to the boil and blanch the whole ears of corn for 5 minutes. Strain the corn from the water and remove the husk, then strip the kernels from the cob using a sharp knife. Set aside in the refrigerator until required.

Now to combine the filling. Into the deep container with the chilled velouté, add the corn kernels and the picked leaves of baby spinach. Combine well, taste and season with salt and pepper if needed. Remove the cold fish pieces from the fridge and drain away any juices. Add the fish to the mixture and gently combine, taking care not to break up the fish too much. Set the mixture aside.

On a clean and cold work surface, lay out the pastry to cut it to size. You will be lining the base and top of these pies, so cut a larger circle to cover the base and come up the sides with 1 cm (½ in) of overhang, and a smaller sheet to cover the top of the pie. For this we use circular 142 ml (4½ fl oz) dishes measuring 9 cm (3½ in) in diameter and 2.1 cm (¾ in) deep.

Once the pastry is cut to size, line the bases of the pie dishes with the larger rounds of pastry. Gently push the pastry into the edges so there is as much space as possible for the filling. Spoon 120 g (4½ oz) of the chilled filling into each pie and spread it out to fill the gaps. Place the smaller round of pastry directly over the top of the filling and crimp the overhanging pastry from the base layer over the top to create a neat seal.

With a pastry brush, paint the tops with the egg yolks, ensuring an even layer.

Place in the refrigerator for 30 minutes to dry the egg glaze. Add another coat of egg wash to the pastry before baking to create a bronzed finish. Bake in a 185°C (365°F) oven for 25–30 minutes.

EMPANADAS

I love these empanadas so much – whether they are cooked from fresh or frozen, they are always consistent and adored by all. See the recipe as a good ratio of ingredients that can be adjusted and played with to suit your own style. From pasties to curry puffs, let this be the starting point for a great pastry and filling recipe.

MAKES 125 EMPANADAS

PASTRY

500 g (1 lb 2 oz) butter
3 kg (6 lb 10 oz) plain (all-purpose) flour
3 tablespoons baking powder
60 g (2 oz) salt
1.25 litres (42 fl oz/5 cups) water

FILLING

150 ml (7 fl oz) grapeseed oil
1.5 kg (3 lb 5 oz) skinless, boneless swordfish mince
2 red capsicums (bell peppers), finely diced, seeds and core removed
2 brown onions, peeled and finely diced
150 g (5½ oz) dried currants
1 litre (34 fl oz/4 cups) fish stock
2 teaspoons dried oregano
30 g (1 oz) sweet paprika
6 g (⅛ oz) ground cumin
chilli flakes, to taste
salt, to taste

To make the pastry, melt the butter in a saucepan over a low heat. Cool to approximately 37°C (100°F). Combine the dry ingredients and form a well in the centre. Add the water and melted butter into the well and slowly incorporate into the dry ingredients. Knead for up to 5 minutes until the pastry is smooth and elastic. Place in the refrigerator overnight or for at least 4–5 hours to rest, so the flour can properly hydrate.

For the filling, in a large cast-iron frying pan, add 50 ml (1¾ fl oz) of the oil and allow it to come to smoking point over a high heat. Add a third of the swordfish mince and, using a whisk, break the mince up. You want to fry the mince quickly with minimal liquid coming off so that it doesn't boil in its own juice and dry out. Remove each batch of browned mince from the pan into a colander set over a bowl. Heat the pan again and cook the remaining two batches of the mince, adding another 50 ml (1¾ fl oz) of oil each time. Set the browned mince aside.

Wipe the pan clean, set over a medium heat and add the remaining 50 ml (1¾ fl oz) of oil, then sweat off the capsicum and onion until softened, approximately 15 minutes, and add the remaining ingredients except the chilli and salt. Bring to a boil then lower the heat to a simmer and allow to reduce by half.

Add the cooked swordfish mince back to the pan and remove from the heat. Season to taste with chilli and salt, then allow to cool.

Roll the pastry out to 4 mm (¼ in) thick sheets using a rolling pin on a lightly floured surface (but not too much flour as this will affect the final exterior appearance of the empanada). Cut into circles using a 12 cm (4¾ in) round pastry cutter.

Place a heaped tablespoon of the cooled filling into the centre of the pastry. Use water to seal the edges shut, pushing out any air pockets first. Crimp the edges with your fingers or use a lightly floured fork to pinch the edges together.

Store in the freezer, ensuring the empanadas are well covered to avoid freezer burn on the pastry.

These empanadas are best cooked from frozen in a deep-fryer set to 180°C (365°F). Fry until golden brown and the filling in the centre is hot. Serve with garlic yoghurt or a good squeeze of lemon.

HOT-SMOKED FISH PIE

This is one of two pie recipes in this book, and while the methods are similar, I absolutely love a smoked fish pie.

Smoking the bones of the fish the day before attempting this recipe gives the stock that is the base for the pie sauce a well-rounded, smoky, savoury profile. One other point here is to commit the trim of fish from the heads, tails and collars that don't often get the full opportunity to shine. Feel free to use seasonal fish or vegetables in place of what's suggested below.

MAKES 12 LARGE PIES

- 3.75 kg (8 lb 4 oz) white fish trim, skinned and boned (a moist, flaky variety such as blue eye trevalla or cod works well)
- 75 g (2¾ oz) salt
- 1.25 kg (2 lb 12 oz) fish bones (any frames and bones from white, clean-flavoured fish work well)
- 4 fresh bay leaves
- 6 thyme sprigs
- 2 tablespoons black peppercorns, coarsely cracked
- 1 tablespoon cloves, coarsely cracked
- 2.5 litres (85 fl oz/10 cups) full-cream (whole) milk
- 2.5 litres (85 fl oz/10 cups) white fish stock
- 480 g (1 lb 1 oz) unsalted butter, diced into 2 cm (¾ in) cubes
- 450 g (1 lb/3 cups) plain (all-purpose) flour
- ⅛ teaspoon ground cayenne pepper
- ⅛ teaspoon freshly grated nutmeg
- salt and cracked black pepper, to taste
- 5 leeks, white part only, split in half and well rinsed
- 625 g (1 lb 6 oz) brown onions, peeled and finely diced
- 1.25 kg (2 lb 12 oz) green peas, cooked fresh or defrosted frozen
- 3 tablespoons tarragon, picked and finely chopped
- 3 tablespoons parsley, picked and finely chopped
- 2 kg (4 lb 6 oz) puff pastry, rolled to 4 mm (⅛ in) thickness and chilled
- 8 egg yolks

Place the fish trim on a tray in a single layer. Evenly sprinkle the 75 g (2¾ oz) of salt over the whole surface of the fillet and place the tray in the refrigerator overnight to firm up.

Set up a smoker to cold smoke and, once ready, add the fish bones. After a minimum of 4 hours, remove the bones and place them into a large container along with the bay leaves, thyme sprigs, peppercorns and cloves. Cover with the milk and leave in the refrigerator overnight to infuse.

The next day, set up the smoker again, this time to hot smoke at 100°C (210°F).

Remove the fish trim from the fridge and pat dry with paper towel. Place into the hot smoker and smoke until the internal temperature reaches 42°C (110°F) (the cooking time will depend on your smoker). Gently remove from the smoker and place into the refrigerator to cool completely. Once cool, flake the fillet into bite-sized pieces and set aside.

Remove the infusing milk from the fridge and strain, discarding the bones and aromatics. Place the infused milk into a large pot along with the fish stock and place over a low heat. Slowly heat the liquid until it comes to a temperature of approximately 70°C (160°F).

Meanwhile, place 450 g (1 lb) of the butter in a saucepan and set over a medium heat. Once the butter has fully melted, add the flour. Stir to combine into a roux and then cook for 2–3 minutes, or until the rawness of the flour is cooked out. Remove from the heat.

Once the milk stock has reached temperature, take the pot off the heat momentarily then gently spoon in the roux. Use a whisk to break up and combine the roux well into the liquid. Once the lumps of roux have been fully incorporated, place the pot back over a medium heat and continue to whisk until the sauce begins to bubble.

Continue whisking to avoid scorching the bottom while you allow it to cook out for a further 5 minutes. Mix in the cayenne and nutmeg and season with salt and cracked black pepper to taste. Transfer the finished sauce to a deep container or tray, cover the surface with plastic wrap to avoid a skin forming and place in the refrigerator to cool.

Slice the washed leeks very finely into half-moons and steam for 3–4 minutes until tender. Cool the leeks thoroughly post-cooking to avoid discolouration.

Place a small frying pan over a medium heat. Add in the remaining butter followed by the diced onion and a pinch of salt. Sweat the onions, stirring occasionally, until they are soft and translucent and any excess moisture has cooked out. Set aside to cool.

Now you are ready to combine the filling ingredients. Into the deep container with the chilled sauce add the leeks, onions, peas, tarragon and parsley. Combine well, taste and season with salt and pepper if needed. Finally, add in the cold fish pieces and gently combine, taking care not to break up the fish too much. Set the mix aside.

On a clean and cold work surface, lay out the pastry to cut it to size. To line the base and the top of these pies you will need a larger size sheet to cover the base, come up the sides and have an overhang, and a smaller sheet to cover the top of the pie. For this recipe we use 12 rectangular 990 ml (33½ fl oz) dishes measuring 17.2 x 13 x 4.3 cm (7 x 5 x 1¾ in).

Once the pastry is cut to size, line the bases of the pie dishes with the larger sheets of pastry. Gently push the pastry into the edges and corners so there is as much space as possible to place the filling. Spoon 600 g (1 lb 5 oz) of the filling into each pie and spread it out evenly. Place the smaller sheet of pastry directly over the top of the filling. Crimp the overhanging pastry from the base layer onto the top to create a neat seal.

In a small mixing bowl, whisk the egg yolks to an even consistency and use a pastry brush to paint the tops, ensuring an even layer. Place in the refrigerator uncovered for 30 minutes to dry the glaze then brush again with the egg yolk mixture before baking. Bake in a 190°C (375°F) oven for 30–35 minutes.

PÂTÉ EN CROÛTE

I'm always one for a challenge and, in this case, I don't feel it gets much more challenging than a pâté en croûte, let alone a pâté en croûte of fish! Fish Butchery executive chef Rebecca Lara's mastery of both classical technique and modern thinking has brought this exceptional slice of wonder to life.

MAKES 1

FILLING

2.5 kg (5½ lb) boneless, skinless white fish trim
75 g (2¾ oz) Fish Bacon (page 174), diced into 5 mm (¼ in) cubes
250 g (9 oz) Cured Fish Fat (page 250), diced into 5 mm (¼ in) cubes
75 g (2¾ oz) cured hearts, spleens, kidneys, livers (optional)
300 g (10½ oz) brown onions, peeled and finely diced
25 g (1 oz) garlic cloves, peeled and finely chopped
10 g (¼ oz) thyme, picked and finely chopped
10 g (¼ oz) rosemary, picked and finely chopped
10 g (¼ oz) finely ground white pepper
25 ml (¾ fl oz) brandy
75 g (2¾ oz) toasted pistachios
12 g (¼ oz) cure #1
sea salt, to taste

PASTRY

860 g (1 lb 14 oz) plain (all-purpose) flour, plus extra for dusting
22 g (¾ oz) fine sea salt
300 g (10½ oz) unsalted butter, diced into 1.5 cm (½ in) cubes
330 ml (11 fl oz) water
12 egg yolks, beaten, for egg wash

JELLY

300 ml (10 fl oz) port
3 thyme sprigs, washed
3 rosemary sprigs, washed
1 teaspoon white peppercorns, cracked
8 g (¼ oz) titanium-grade gelatine leaves
ice-cold water, for soaking

Begin by placing the fish trim in the freezer to reach a temperature of 0°C (30°F) before you mince. Meanwhile, prepare the rest of the ingredients for the filling and set up a meat grinder with a 10 mm (½ in) plate.

Once the fish has chilled enough, pass it through the grinder into a large mixing bowl. Combine the mince with the rest of the filling ingredients and mix well. Taste and season with sea salt as needed then set aside in the refrigerator.

To make the pastry, place the flour and salt into a Robot-Coupe food processor. Melt the butter gently in a small saucepan over a low heat. Add the water and immediately turn up to the highest heat. You want the water to come to a boil quickly so it doesn't evaporate too much and alter the result.

Remove the hot butter and water as soon as it has come to a boil and, with the Robot-Coupe running, slowly pour into the flour and salt. Continue processing for another couple of seconds until the dough begins to form into a ball, then tip it out onto a clean bench. Knead by hand for another 2–3 minutes. Cover the dough with plastic wrap and allow to rest for 5 minutes to cool only slightly.

Once the dough has briefly rested, place it onto a lightly floured bench and roll it out to a large sheet 7 mm (¼ in) in thickness. You will be cutting the pastry into pieces to fit into the size of your mould – we use a traditional rectangular pâte en croûte mould that measures 30 x 8 x 8 cm (12 x 3¼ x 3¼ in). From your rolled pastry, cut the following pieces: 1 x large rectangle to line the base and the two long sides of the mould, with an approximately 2.5 cm (1 in) overhang; 2 x small rectangles to line the shorter, smaller edges with a 2.5 cm (1 in) overhang; 1 x long rectangle cut to the same size of the mould for the top; and, from the trim, any decorative shapes you'd like. From the long rectangle for the top, punch out three 1.5 cm (½ in) holes that will allow the steam to escape while cooking.

Once the pastry has been cut, you are ready to assemble. Working quickly, because the pastry becomes less pliable the more it cools, lightly spray or brush the mould with a neutral-flavoured oil. Drape the largest rectangle across the base and the longer sides, gently pressing the pastry into the corners and edges of the mould. With the yolks, brush a little egg as glue, then press in the two smaller rectangles to line the short sides.

Remove your filling from the fridge and begin to press into the pastry-lined mould. Fill the mould little by little, making sure no air pockets are forming and being gentle so as not to damage the pastry lining. Once all the filling has been packed into the pastry, cover with the top, hole-punched piece of pastry.

Brush the edges of this top pastry with egg yolk as glue then crimp over the overhanging pastry, ensuring the filling is completely sealed off. Use the yolks as a glue to stick on any decorative pieces you have cut.

With a pastry brush, paint the top with a thin, even layer of egg yolks, avoiding any thick clumping, then place in the refrigerator. Repeat this process until the pâté en croûte is glazed with a total of three layers of egg yolks, placing in the refrigerator for approximately 30 minutes between glazes to dry before layering on the next.

Preheat a combination steam and convection oven to 200°C (390°F).

Once the final glaze on your pâté en croûte has dried, use a sharp paring knife to lightly score the pastry lid as desired; this adds to the decoration of the pâté as well as allowing the glaze some room to crack in a way that you prefer. Using a folded piece of foil, fashion three small cylinders 3–4 cm (1¼–1½ in) in length and as wide in diameter as the punched holes in the lid. Gently push them into the holes to assist the steam to escape during baking.

Place the pâté en croûte on a baking tray and into the preheated oven.

Cook until a skewer inserted into the centre of the pâté comes out feeling warm. Keep a close eye on it to ensure the pastry reaches a dark golden colour without overcooking the filling. (You may need to increase the temperature to speed up the caramelisation if you find the filling is close to finishing before you have achieved the colour you'd like.) This cooking process will take 20–30 minutes, depending on the size and shape of your mould.

Once cooked, remove the pâté from the oven, cool for 10 minutes at room temperature, then transfer to the refrigerator and chill overnight. It is important to allow the pâté en croûte time to properly chill as you want the filling to shrink and set, creating gaps for the jelly.

The next day, prepare the jelly. Place the port, thyme and rosemary sprigs and cracked white peppercorns into a saucepan, place over a medium-high heat and bring to a boil. Allow the alcohol to cook out of the port and the mixture to reduce by half.

Meanwhile, soak the gelatine leaves in the ice-cold water until they have fully softened.

Once the port reduction is ready, strain off and discard the aromatics. Squeeze out the excess water from the softened gelatine leaves, then add to the hot reduction and stir to dissolve fully. Strain the mixture to ensure there are no clumps of undissolved gelatine.

Pour into a small spouted jug and set the jelly aside to cool slightly to blood temperature.

Remove the chilled pâté en croûte from the fridge. Using the foil chimneys as funnels, slowly begin to pour in the liquid jelly. The amount of jelly that you will need differs each time, so ensure you pour slowly as you don't want it to overflow over the top.

Allow it to sit at room temperature for a further 10–15 minutes as the jelly might take time to find its way into all the gaps inside. You want to ensure that all the gaps have been completely filled with the jelly before it sets. It is full when the chimneys are visibly full of jelly

Once you are satisfied with the level of jelly, gently place it into the refrigerator and allow to set for a minimum of 3 hours. To serve, gently remove from the mould onto a cutting board and slice with a sharp serrated knife.

TUNA WELLINGTON

The tuna wellington was a much-loved recipe that we developed during Saint Peter's Covid closure in 2020. Having demonstrated its technical challenges on *MasterChef Australia*, it is still a dish that I am very proud of and one that regularly appears on both the menu at Saint Peter and as a part of the retail offering at Fish Butchery.

MAKES 2 SERVES

1 x 300 g (10½ oz) tuna loin, approximately 12 x 5 x 5 cm (4¾ x 2 x 2 in)
salt and pepper, to taste
3 egg yolks, beaten, for egg wash

MUSHROOM DUXELLES

1 kg (2 lb 3 oz) portobello mushrooms
300 g (10½ oz) ghee
3 brown onions, finely diced
1 head of garlic, peeled and grated
1 bunch of thyme, picked
salt flakes and freshly cracked black pepper, to taste

CREPES

75 g (2¾ oz) butter
60 g (2 oz) plain (all-purpose) flour
60 g (2 oz) buckwheat flour
5 g (⅛ oz) salt
2 eggs
300 ml (10 fl oz) hoppy beer

SOUR CREAM PASTRY

400 g (14 oz) plain (all-purpose) flour, plus extra for dusting
10 g (¼ oz) fine salt
280 g (10 oz) cold butter
200 g (7 oz) sour cream
20 g (¾ oz) ice-cold water

For the mushroom duxelles, preheat the oven to 220°C (430°F). Roughly break the mushrooms by hand into quarters.

Place 100 g (3½ oz) of the ghee into each of two baking trays and place in the oven to preheat. After 3–4 minutes, when the ghee is shimmering, add the mushrooms to the trays and season well with salt flakes and freshly cracked black pepper. Roast the mushrooms until well caramelised and tender, approximately 10–15 minutes, then drain in a colander.

In a heavy-based frying pan, heat the remaining 100 g (3½ oz) of ghee to a light haze over a medium heat. Add the onions, garlic and thyme and sauté for 5–6 minutes until the onions and garlic begin to take light, even colour.

Add the mushrooms to the frying pan and cook over a medium heat until all the juices have reduced and the onions and mushrooms begin to catch, approximately 10 minutes. Be careful at this point as the mixture is very easy to burn.

Remove from the heat and, using a perforated spoon to drain the mushrooms further, finely chop the hot mushrooms in a food processor but do not puree. Spread onto a tray and chill.

For the crepes, melt the butter in a saucepan over a low heat, or microwave. Sift together the plain flour, buckwheat flour and salt into a large mixing bowl. In a clean bowl, crack the eggs and lightly whisk together to make an even liquid. Make a well in the flour and add the eggs, then the beer and then the butter, using a whisk to combine each liquid as it is added. The mixture should be free of any lumps. Leave to rest for a minimum of 30 minutes before cooking.

Lightly grease a cast-iron pan with a splash of neutral cooking oil (grapeseed or canola) and heat over a medium flame. When the pan is hot, add a thin layer of the batter and swirl around until the mix just touches the edges. When the crepe is lightly toasted on the bottom, turn the crepe over and cook briefly for 20 seconds until the batter is cooked through. Turn out onto lightly greased baking paper and repeat for a second crepe.

For the sour cream pastry, pulse the flour, salt and butter together in a food processor, being careful not to blend too far as coarse pieces of butter through this pastry are important for its final texture. Add the sour cream and chilled water to the food processor and pulse again. The mixture should still be crumbly but come together in your hand. Work with your palms on the bench till the dough just comes together, no more. (You should be able to see the pieces of butter and ripples of sour cream.) Refrigerate for at least 15 minutes before using.

Construction

Roll the pastry out on a lightly floured work surface to a 3 mm (⅛ in) thick sheet approximately 28 cm (11 in) wide and 45 cm (18 in) long. Cover and leave to rest on a baking tray in the refrigerator.

Line a bench with plastic wrap twice the size of the two crepes and lay them out on it, slightly overlapping each other going away from you. Spread the mushroom duxelles out on the crepes to 5 mm (¼ in) thickness and just a little longer than the length of the tuna.

Season the tuna loin well with salt and pepper. Place in the centre of the crepes and lift the bottom crepe over the tuna to wrap it in duxelles, then roll the tuna over so it's enclosed in the duxelles and crepe. (If you have too much excess, trim the crepes.) Fold down the end so it's just covering, then wrap tightly with the plastic wrap to form a cylinder. Tie the ends of the plastic firmly in a knot and place in the fridge to chill and set for 30 minutes.

Place the pastry on a bench and brush with egg wash. Carefully unwrap the tuna and work out where you need to place it to wrap it in the pastry with minimal overhang, leaving 5 cm (2 in) at each end to cover. Then roll the tuna until it's wrapped just once and cut along the join, leaving 5 mm (¼ in) of extra pastry and pressing together so the seam becomes invisible.

This seam is your base, so put it on the bottom. Pull the pastry down at each end, then fold the ends in as if you were wrapping a present.

Place the wellington on a baking paper–lined cast-iron tray. Brush with egg wash, then leave in the fridge for 10 minutes before brushing again.

Preheat the oven to 200°C (390°F) then bake for 10–15 minutes. The internal temperature should be approximately 25°C (75°F) when removed from the oven. Leave to rest for 5 minutes before carving. Once rested, the tuna will be approximately 38°C (105°F). Using a sharp chef's knife, slice thick portions from the wellington and serve immediately.

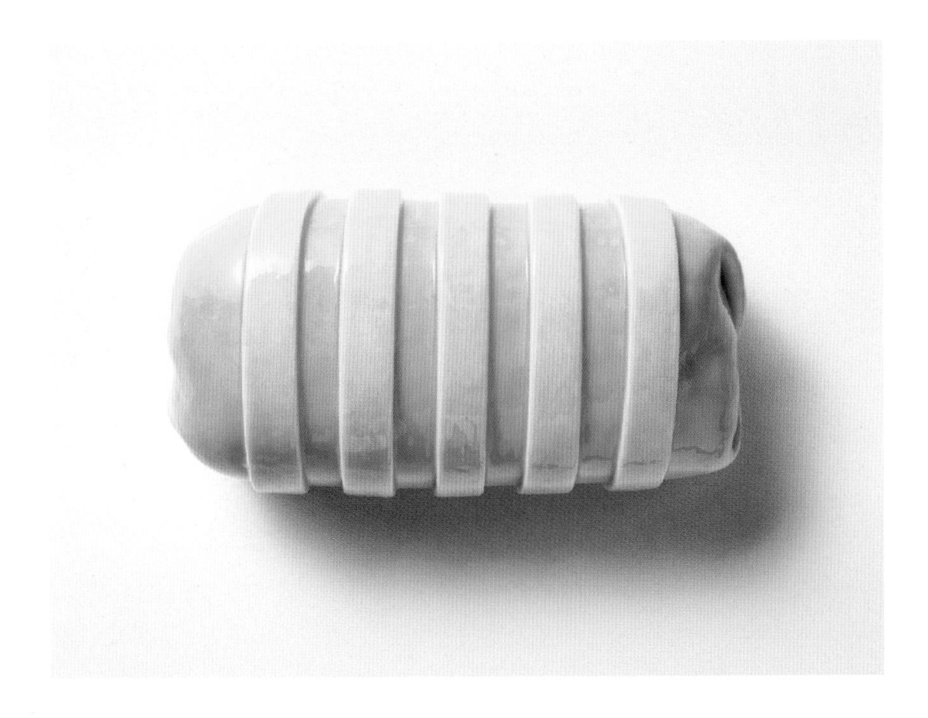

SCHNITZELS

Who doesn't love a good schnitty? Or, at the very least, something crumbed and fried to golden perfection. Crumbing any fish fillet removes a layer of anxiety from the challenges of cooking fish at home. There is something about the ability of a crumbed coating to insulate and protect the flesh from the harshness and oftentimes unforgiving nature of direct contact with a pan or grill.

Mix up the species of fish you crumb and the seasonings you add to the crumbs and flour.

MAKES 2 SCHNITZELS

500 g (1 lb 2 oz) boneless fish fillet, skin on
2 large eggs
50 g (1¾ oz) good-quality parmesan cheese, finely grated
50 g (1¾ oz) dried oregano
1 teaspoon dried chilli flakes
180 g (6½ oz/3 cups) white panko breadcrumbs
250 g (9 oz/1⅔ cups) plain (all-purpose) flour
1 tablespoon smoked paprika
1 tablespoon finely chopped parsley (optional)
75 g (3½ oz) ghee
sea salt flakes and freshly cracked black pepper

Take a sharp knife and, working on a steep angle, cut two thickish medallions from the fillet of fish. (By cutting the fish on an angle you create a greater surface area, so the schnitzel will cook far more evenly.) The portions should be approximately 2 cm (¾ in) thick.

Place the eggs in a shallow bowl and whisk to combine.

Add the parmesan, oregano and chilli to the breadcrumbs and then add the paprika to the flour. Tip the flour and breadcrumbs onto separate flat baking trays ready to crumb the fish.

Dip a fish medallion first into the flour to coat all sides, clapping away any excess, then dip it into the egg mixture, allowing the majority of the egg to drip away.

Lastly, coat with the breadcrumbs, pushing down firmly so the breadcrumbs stick evenly from edge to edge. Scatter over the parsley, if using.

Melt the ghee in a large cast-iron frying pan over a medium heat. Add the fillets skin side down to the pan and cook for approximately 2 minutes on each side, keeping the pan moving to swirl the fish around in the hot fat and turning it over halfway through cooking until golden and evenly coloured on both sides. Season liberally with salt and pepper, then transfer to a plate lined with paper towel to rest before serving.

FISH FINGERS

I grew up eating fish fingers from a box, not having any idea where they came from or what fish they were from – and let's be honest, when you're six years old, these aren't the big questions you ask yourself. However now, as a father of four, I want to introduce the idea of knowing where your food is from and from what sort of fish or animal it was. This knowledge empowers the consumer to see the true value of the product while having greater confidence that this is something they want to feed their children. It might be difficult to sell a fish head over the counter, but apply some labour to picking the meat away from the bones and you have a product that is a wonderful alternative to the fish fingers we grew up eating.

MAKES 70–80 FISH FINGERS

3 kg (6 lb 10 oz) picked head and collar meat (preferably a gelatine- and fat-rich fish like cod, monkfish or coral trout)
50 g (1¾ oz) salt
15 g (½ oz) ground black pepper
10 g (¼ oz) ground fennel seeds
500 g (1 lb 2 oz/3⅓ cups) plain (all-purpose) flour
12 eggs, beaten
500 g (1 lb 2 oz) panko breadcrumbs

To make the mix for the fish fingers, ensure you start with warm, freshly picked fish head and collar meat (see page 60), then combine with the salt, ground pepper and ground fennel seeds.

Use a large sheet of plastic wrap to line a 53 x 32.5 x 1 cm (21 x 13 x ½ in) baking tray. Ensure you have a little bit of excess wrap hanging over the two short edges as this will help you remove the set filling from the tray.

Tip the seasoned head and collar meat into the lined tray and spread out evenly, smoothing any bumps and pressing out any pockets of air. Cover with another sheet of plastic wrap and use an identical tray to press the mixture down. Fill this tray with something heavy to evenly weigh down the mixture and place the whole thing into the refrigerator for a minimum of 3–4 hours, or until completely firm and set.

Once set, remove the weighted tray and peel away the top sheet of plastic from the set mixture. Invert the whole tray onto a cutting board and, using the overhanging plastic, gently pull the mixture out of the tray until it falls flat onto the board.

Once it has come loose, remove the tray and the remaining plastic sheet.

Using a clean, sharp knife, cut the slab into three long pieces along the shorter end of the tray, then portion into 2 cm (¾ in) wide fingers. You should end up with fingers approximately 10 x 2 x 2.5 cm (4 x ¾ x 1 in). Feel free to square off the edges first if you prefer all the fingers to be perfectly exact in size (this trim can be utilised as head meat in other recipes, such as Fishcakes on page 221).

Set up a crumbing station by arranging the flour, eggs and breadcrumbs, in order, in three wide-surfaced, shallow containers.

Crumb the fish fingers by dipping them first into the flour then into the eggs and lastly into the breadcrumbs. Gently press the breadcrumbs into the fish fingers to help them adhere to the egg, ensuring an even coating. Arrange the finished fish fingers onto a tray and refrigerate or freeze until ready to use. To cook, the fish fingers can either be shallow-fried in ghee in a cast-iron pan set over a medium heat, deep-fried at 180°C (360°F), or baked in a preheated 190°C (375°F) oven for approximately 20 minutes, or until golden brown.

FISHCAKES

Time to give the humble fishcake the attention it deserves! So much fish meat goes to waste, especially from the heads, collars, tail and meat that is left on the bone after filleting, and it would be good to see this fish find a home. There are few better vehicles for moving this type of fish than adding some beautiful herbs, potatoes and breadcrumbs. Once assembled, these fishcakes are easy to pan-fry, and everyone from young to old will enjoy them. Investing time and labour in removing the bones and using up the trim is not only economically sensible but also ethically responsible.

MAKES 30 FISHCAKES

2.5 kg (5½ lb) potatoes (a floury variety such as sebago or kestral works well)
1 tablespoon extra-virgin olive oil
500 g (1 lb 2 oz) brown onions, peeled and finely diced
2 kg (4 lb 6 oz) picked head and collar meat (page 60)
3 tablespoons finely chopped flat-leaf (Italian) parsley leaves
3 tablespoons finely chopped dill leaves
3 tablespoons finely snipped chives
zest of 3 lemons
1 tablespoon ground fennel seeds
500 g (1 lb 2 oz/3⅓ cups) plain (all-purpose) flour
12 eggs, beaten
500 g (1 lb 2 oz) panko breadcrumbs
salt and pepper, to taste

Preheat a combination steam and convection oven to 100°C (210°F).

Place the cleaned whole potatoes in a single layer on a perforated tray and steam in the preheated oven until tender. Start by checking them after 30 minutes by piercing with a paring knife; they may take longer depending on the potatoes you are using. Once cooked, remove the tray and, using a tea towel (dish towel) and paring knife to handle the hot potatoes, peel away the skin. Push the skinless potatoes through a potato ricer into a large mixing bowl and set aside.

Place a small frying pan over a medium heat. Add the olive oil followed by the diced onion and a pinch of salt. Sweat the onions, stirring occasionally, until soft and translucent and any excess moisture has cooked out. Spoon the cooked onions on top of the mashed potatoes and set aside to cool.

Once the potatoes and onions have cooled to room temperature, add the cooked meat from the head and collars to the bowl along with the chopped herbs, lemon zest and ground fennel seeds.

Wearing gloves, use your hands to combine the mixture together, ensuring to break up and evenly disperse any flakes of meat that may have gelled together while in refrigeration. Taste the mixture and season with salt and pepper to your preference.

Once you are happy with the taste of your fishcake mixture, portion them for crumbing. With a weighing scale and gloves, use your hands to scoop up and weigh out into 140 g (5 oz) portions. Shape each into a fishcake approximately 9 cm (3½ in) in diameter and 1.5 cm (½ in) thick. Place the fishcakes on trays, cover and refrigerate for 30–60 minutes to chill and firm up.

Set up a crumbing station by arranging the flour, eggs and breadcrumbs in order, in three wide-surfaced, shallow containers.

Remove the fishcakes from the fridge and dip them first into the flour, then into the eggs and lastly into the breadcrumbs. Gently press the breadcrumbs into the fishcake to help them adhere to the egg, ensuring an even coating. Arrange the finished fishcakes onto a tray and refrigerate until ready to cook. The fishcakes can either be shallow-fried in ghee in a cast-iron pan set over a medium heat, deep-fried at 180°C (360°F), or baked in a preheated 190°C (375°F) oven for approximately 35 minutes, or until golden brown.

TUNA PATTIES

I grew up eating tuna rissoles made from canned tuna, mashed potatoes and breadcrumbs, but these are a little different.

Tuna patties made from ground tuna trim are an integral part of the double yellowfin tuna cheeseburger that we serve at Fish Butchery and Charcoal Fish. The ambition for this burger is not to showcase the delicate nuances of the tuna but to step into the idea that all the sinews and aesthetically compromised pieces of the fish can be as meaty as beef.

This recipe can be extended to tuna meatballs, koftas or even meatloaf.

MAKES 75 PATTIES

- 5 kg (11 lb) boneless, skinless tuna trim
- 25 g (1 oz) table salt
- 15 g (½ oz) ground black pepper
- 8 g (¼ oz) ground fennel seeds
- 45 g (1½ oz/1½ packed cups finely chopped parsley leaves

Set up a commercial meat grinder with a 13 mm (½ in) plate. Place the tuna in the freezer. Once below 0°C (30°F), mince through the grinder.

Combine the ground tuna with all of the remaining ingredients in a large mixing bowl. Form a small patty and cook it in a hot frying pan to check the seasoning.

Lay two sheets of plastic wrap approximately 60 cm (23½ in) long along a bench and place a 2.5 kg (5½ lb) log of the mixture along the bottom of each leaving about 5 cm (2 in) of space at each end. Wrap each log tightly in the plastic wrap by holding the ends and rolling to a length of 50 cm (19¾ in) long. Tie the plastic off at both ends with a knot. Place the logs in the freezer and turn every hour or so until set so the result is cylindrical and not flat on one side.

Using a sharp knife, slice the frozen logs into 70 g (2½ oz) patties that are approximately 1.5 cm (½ in) thick. The patties are best cooked from frozen over a charcoal grill or on a gas barbecue to achieve maximum colour while not drying out the fish. Brush first with a little cooking oil and then season lightly with salt. Ensure the grill is on high or that the coals are pushed together to create a high heat. Grill for about 2 minutes on each side and ensure they are well caramelised to generate plenty of meaty characteristics. Serve on burger buns with all your favourite toppings.

SNACKS AND SUNDRIES

ROE TO CAVIAR

Sturgeon caviar is regarded as one of the most desirable and luxurious food products on the planet. In this recipe I attempt to bring the same desirability to the roe of a fish that isn't a sturgeon. I have tested this recipe on the roe of Murray cod, John Dory, mirror dory, blue eye and hapuka. Across all these species I achieved a firm popping texture on the eggs along with a moreish level of seasoning that makes you want another spoonful.

MAKES 100 G (3½ OZ)

2 whole fresh fish roe (approximately 150–200 g (5½–7 oz) each to yield 130 g (4½ oz) scraped, cleaned eggs)
6 g (⅛ oz) table salt, depending on the roe

Start by cutting the membrane of the roe sac to expose the eggs within. Using the same knife, scrape the eggs out of the membrane and place in a large bowl of ice-cold water.

Using a whisk, separate the eggs from the loose bits of membrane that reside within the roe – what attaches itself to the whisk is what is removed. This will take several attempts and it will feel like you are discarding quite a bit, but there is a considerable amount of membrane within the roe. Once the eggs are loose and completely separated, drain them using a sieve and discard the water.

Transfer the eggs to a cloth or towel to drain really well; you may need to change these a few times. Once dried, place the roe in a bowl, add the salt and stir together. Stand for 10 minutes to allow the salt to dissolve.

Drain the roe once more and leave the eggs exposed on a tray to dry in the refrigerator for a minimum of 4–5 hours.

Once the eggs reach a consistency that is firm but yielding and the clarity of the eggs is noticeably more refined than before salting, store the eggs either in a tin or airtight container. Serve generously with oysters or where sturgeon caviar would be used.

CURED ROE

This is not technically a bottarga and nor is it trying to be – this is our version of a cured roe that is lightly smoked and left with some chewiness and waxiness, which I absolutely love about this product. If you don't enjoy this texture, dry the roe further for one that is firm enough to thinly slice or grate over dishes.

MAKES 750 G (1 LB 11 OZ)

100 g (3½ oz) light brown sugar
300 g (10½ oz) fine salt
1 large, firm and intact roe sac from a large fish such as hapuka or blue eye trevalla
cure #2

To cure the roe, stir together a curing mix of one-part light brown sugar to three-parts salt in a mixing bowl.

Toss the roe gently (so as not to break it) with the curing mix along with 2 g (⅛ oz) of cure #2 for every 1 kg (2 lb 3 oz) of roe. The roe should be coated as you would coat something in breadcrumbs. Place the roe into a vacuum-pack bag with any loose curing mix and seal tightly in a cryovac machine. Refrigerate for at least 1 week, turning regularly.

Once the roe has firmed up significantly, it is ready to smoke. Remove from the vacuum bag and wipe off any excess surface salt and sugar. In a smoker, cold smoke for 2 hours. Remove from the smoker, transfer to a wire rack set over a baking tray and refrigerate uncovered until completely cold.

Once chilled, use butcher's twine to tie around the top of the roe to create a loop and hang on a butcher's hook in a fan-forced refrigerator for at least 3 weeks to firm up.

After 3 weeks, the roe should be firm to the touch but still have a slight spring. This will give the roe a nice waxiness as opposed to something that is dry or even crumbly. Slice the roe thinly and serve in some good extra-virgin olive oil, like an anchovy.

MAW CRACKLING

This is a fantastic method that brings textural life to an otherwise challenging part of the fish. See it as a pork crackling that's best enjoyed with a beer and plenty of salt.

MAKES 1 LARGE CRACKLING

100 g (3½ oz) fish maw
grapeseed, canola or cottonseed oil, for deep-frying
sea salt flakes

Cut the maw down one side, then lay out flat. Using a pastry card or knife, carefully scrape away any imperfections, then transfer to a saucepan and cover with cold water. Bring to the boil, then reduce the heat to a very gentle simmer and cook for 30 minutes, or until the maw is soft, translucent and almost jelly-like in consistency.

Meanwhile, preheat the oven to its lowest setting, about 60–70°C (140–160°F). Line a baking tray with baking paper.

Remove the maw from the liquid with a slotted spoon and spread out on the prepared tray. Transfer the tray to the preheated oven and leave until completely dry, up to 5 hours. At this point the maw can be stored indefinitely in an airtight container for later use.

To make the maw crackling, half-fill a large saucepan with oil for deep-frying and heat over a medium–high heat to 185–190°C (365–375°F). Using a small pair of tongs, very carefully place the dried maw in the oil for just a few seconds until the skin has puffed up and tripled in volume but not coloured. Quickly remove from the oil and drain on a wire rack.

Season liberally with salt then serve as a snack or cut into smaller pieces to garnish a dish of raw fish for additional texture and flavour.

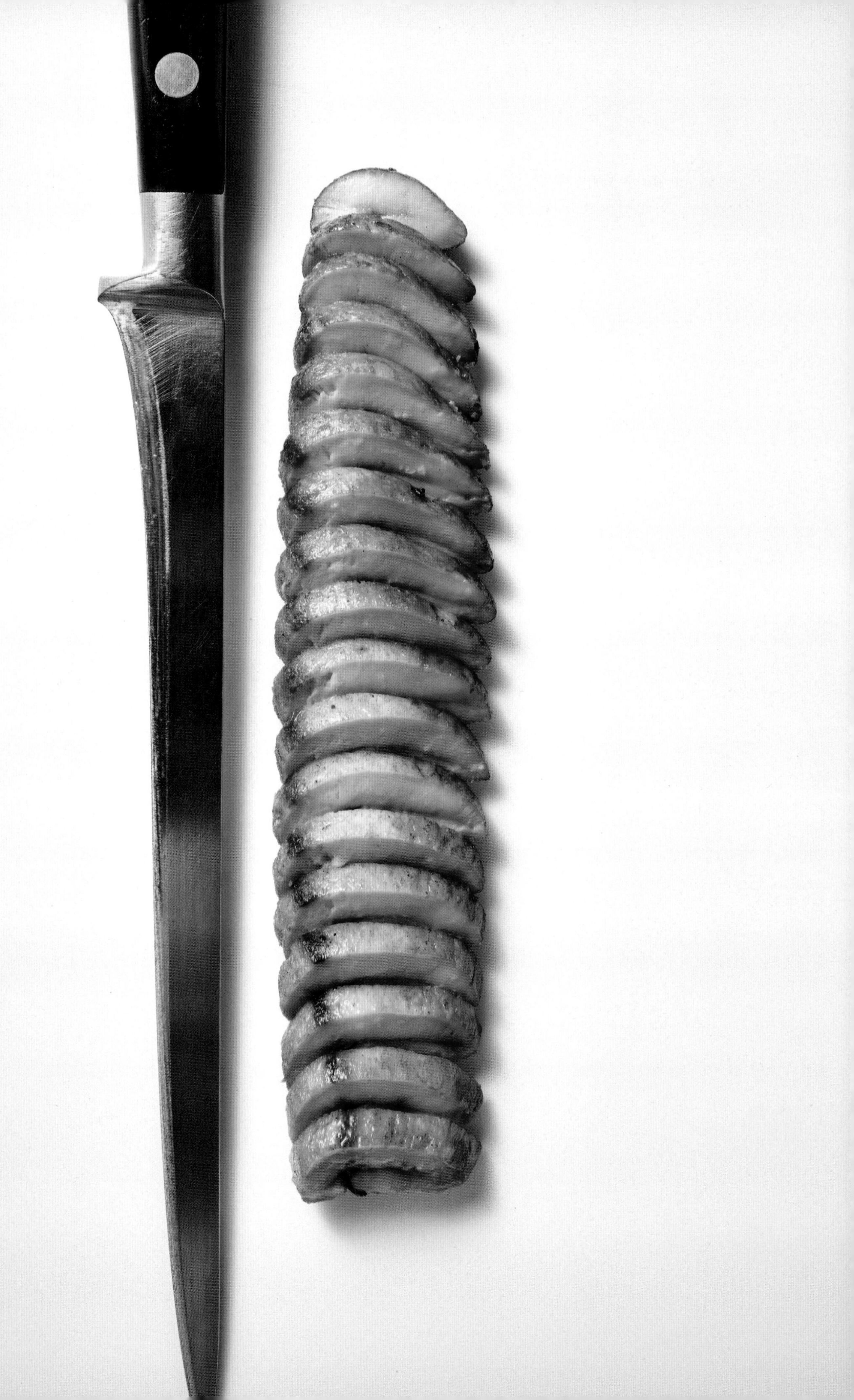

SMOKED MILT

Milt (aka fish sperm) can provoke much hesitation in Western cookery due in part to its texture and appearance. However, if you start to see it more as meat offal, like brains or sweetbreads, the opportunities open up. This method is assistive as it again preserves the offal, meaning there is less of a ticking clock for it to be consumed. The curing, smoking and drying are all critical variables to achieve a really delicious outcome.

My favourite way to work with this product is to pan-fry slices of the smoked milt in ghee and serve it with the fish that it was from or on toast, or even diced and added to the mortadella, sausage or terrine recipes.

MAKES 1

1 large, firm and intact milt sac from a large fish such as hapuka or blue eye trevalla

CURE

(use 130 g (4½ oz) per kilogram (2 lb 3 oz) of milt)

250 g (9 oz) caster (superfine) sugar
500 g (1 lb 2 oz) table salt
75 g (2¾ oz) ground cumin
25 g (1 oz) cure #1
75 g (2¾ oz) ground black pepper
75 g (2¾ oz) ground coriander seeds
75 g (2¾ oz) ground fennel seeds

For the milt cure, add all ingredients to a clean stainless steel bowl and combine. Once combined, keep in a clean, dry container for future use.

Weigh the milt sac and work out the amount of cure that you will need.

Coat the milt sac evenly in the cure mix as you would coat something in breadcrumbs. Place uncovered on a clean stainless steel tray and refrigerate for at least 2 days, depending on its size, turning regularly.

Once the milt has firmed and it has dropped a lot of moisture, it is ready to smoke. Remove from the cure and wipe off any excess.

In a smoker, cold smoke the milt sac for 2 hours. Place uncovered on a perforated tray and allow to dry in a refrigerator for at least 48 hours. The ideal texture of the smoked milt will be a slightly leather-like dry finish around the outside with a firm but buttery interior. Alternatively, the milt can be dried out completely and then ground to use as a seasoning.

SALT AND VINEGAR FISH

At Saint Peter, I always announce this as one of my favourite dishes to eat. Be sure to pour gordal olive brine over the pickled fish and have some warm sourdough and cultured butter at the ready. This is a wonderful method for any fish: just adjust the salting and vinegar times based on the size of the fish, but this example is a great starting point. Only leaving it in the vinegar briefly maintains a slight rawness in the centre of the fillet, which gives an amazing final texture that doesn't taste as familiar as pickled fish.

MAKES 4

4 very fresh blue mackerel fillets, ribs removed, tail intact (about 80 g/2¾ oz each)
80 g (2¾ oz/⅔ cup) sea salt flakes
250 ml (8½ fl oz/1 cup) champagne vinegar

To pickle the mackerel fillets, leave the pin bones in place. Season both the skin and flesh sides evenly with the salt, then place on a tray and refrigerate, uncovered, for 1½ hours.

Once that time has passed, rinse the salt from the fish in a small amount of the vinegar. Submerge the vinegar-rinsed fish in a bath of the remaining vinegar and leave to pickle for 20–25 minutes.

Remove the pickled mackerel, reserving the vinegar for the next batch you make. Remove the pin bones from the fish at this stage using pliers.

Turn the fish onto the flesh side so the skin is facing up. Using a pair of fish pliers or tweezers, grab onto the corner of the skin closest to where the head would have been and pull the skin gently off the flesh, leaving behind the silver skin.

Slice the mackerel about 5 mm (¼ in) thick, working from the head to the tail.

Assemble a mackerel across each of four serving plates.

OLIVE OIL–COOKED TUNA

I enjoy cooking in olive oil as it brings such a unique texture and flavour to the final outcome. Texturally, I find canned tuna to often be far too dry and lacking in seasoning, so utilising a very sinew-heavy cut of tuna in this way transforms it into something remarkable for salads and sandwiches. The cut we selected (or, rather, the cut that was constantly pushed aside) is the sinew-rich chain that sits adjacent to the dorsal fins of the fish. However, the tougher cut from the tail end of the tuna or even the very fatty belly will also give a great result.

Have a play around with different flavours to find something you really enjoy.

SERVES 4

1 skinless tuna chain cut, about 1.5–3 kg (3 lb 5 oz–6 lb 10 oz)
table salt
peeled zest of 1 lemon
2 dried red chillies
1 fresh bay leaf
1 bunch of rosemary
1.5 litres (51 fl oz/6 cups) extra-virgin olive oil

For the tuna, use a sharp knife to cut the length of the chain into thirds.

Place the tuna chain on a tray, season liberally with salt and lay the lemon zest, broken dried chillies, bay leaf and the whole rosemary sprigs over it. Leave uncovered in the refrigerator for approximately 3 hours.

After this time has passed, place the seasoned tuna into a deep baking dish with a wire rack set inside it. Pour over the olive oil and cover the dish with a square of baking paper.

Place in a low oven set to 65°C (150°F) and cook the tuna until it reaches an internal temperature of 58°C (135°F).

Remove the tray from the oven and keep the tuna fully submerged in the oil until it is completely cooled.

Transfer to the refrigerator once cooled, keeping the tuna submerged to gain flavour from the fragrant oil. The tuna can be cut into smaller pieces, covered with the oil and stored in vacuum-pack bags sealed in a cryovac machine or in a sterile glass jar in the refrigerator.

Use this cooked tuna through salads, on toast or blended into a fantastic tonnato sauce.

GARFISH STICKS

These skewers are a thought-provoking and visually striking preparation that could be the catalyst for many other interpretations. As a consumer, knowing these are boneless gives so much confidence to enjoy this beautiful fish straight off the skewer.

While they can be marinated or further vegetables added to the skewer, there is something beautiful about the fish in this simple preparation. And this is one of my favourite dressings, so be sure to squirrel it away as it works on everything!

MAKES 6

6 boneless, butterflied garfish
60 ml (2 fl oz/¼ cup) grapeseed oil
sea salt flakes
6 metal or soaked bamboo skewers

DRESSING

50 g (1¾ oz) achiote paste (available from a Mexican grocer)
250 g (9 oz/1 cup) unsalted butter
80 ml (2¾ fl oz) sherry vinegar
25 ml (1 fl oz) mushroom soy sauce

Lay a butterflied garfish on a cutting board with the head to the left and tail to the right. Make a cut just behind the head to remove the head and collars together. Cut the collars off the head and set aside for another purpose or for stock. Keep the head to one side ready for assembly.

Cut index finger–width squares of the butterflied fillets until you reach the tail. You will have approximately six or seven stacked squares of fillet when finished, depending on the size of the fish.

Taking a sharp skewer, thread the tail on and push down to the halfway point.

Then add each stacked square of fillet to the skewer in the order that they were cut, finishing with the head right at the end of the skewer. Repeat with the other garfish.

Chill the skewers for 10–15 minutes to cool the fish down after having worked with it as it can take time to thread the fillets on carefully.

While you are waiting, add the achiote to a food processor and pulse to a fine crumb. Add to the butter in a large saucepan and set over a medium heat. When the butter is bubbling, whisk the achiote to ensure it doesn't stick to the base of the pan or become lumpy.

After 7–8 minutes, the milk solids in the butter will begin to change to a nut-brown colour and have a toasted hazelnut–like aroma. Remove from the heat and add the sherry vinegar, being careful that it doesn't overflow. Once the pan has settled, add the mushroom soy and keep warm to serve. (Don't strain the achiote – the coarse crumble is chewy and absolutely delicious.)

Preheat a charcoal or gas grill.

Brush the skewers with a little grapeseed oil and season well with sea salt flakes. Place the skewers over a medium heat and, without wanting to apply any colour, allow the smoke from the coals to gently cook the flesh of the garfish. (This will take approximately 4–5minutes, depending on the heat of the coals – you can always cook a little more so please leave some life in the garfish as they don't like too much heat.) Spoon the warm achiote dressing over the garfish and serve immediately.

TUNA BOLOGNESE (RAGU)

While it's common to see meat-based ragus appearing on tables all over the world along with everyone's favourite, spag bol, I truly hope that inside the next five years, ground tuna trim and scraps will be available to buy over the counter in both markets and stores so this meat-based dish can have a fish makeover. This doesn't mean taking more tuna out of the water – it means seeing to it that every last scrap of all the fish that are caught is consumed before the next fish is taken. Don't limit the potential of this ragu to only pasta – this is also a great addition to mashed potatoes, polenta, rice or loaded into a toasted sandwich.

MAKES 5 KG (11 LB)

1.5 kg (3 lb 5 oz) boneless, skinless tuna trim
1 kg (2 lb 3 oz) brown onions, peeled and chopped
750 g (1 lb 11 oz) carrots, peeled and chopped
4 celery stalks, tops removed, chopped
75 g (2¾ oz) garlic cloves, peeled and minced
400 ml (12 fl oz) grapeseed oil
300 ml (10 fl oz) white wine
180 g (6½ oz) jar of tomato paste (concentrated puree)
2.5 kg (5½ lb) tinned crushed tomatoes
1 fresh bay leaf
pinch of freshly grated nutmeg
1.5 litres (51 fl oz/6 cups) water
salt and pepper, to taste

Place the tuna in the freezer. Once below 0°C (30°F), mince through a 2 mm (⅛ in) plate. Keep in the refrigerator until required.

Combine all the chopped vegetables and garlic together in a bowl. In batches, pulse the vegetables together in a Robot-Coupe or food processor to finely chop.

In a large pan set over a high heat, add 150 ml (5 fl oz) of the grapeseed oil and allow it to begin to smoke. Add half the tuna mince to the pan and use a whisk to break it apart so that all the mince separates. (You want to fry the mince quickly with minimal liquid coming off it and maximum colour achieved, so that the mince doesn't boil in its own juices and dry out.) Remove the mince and set aside in a colander to drain off the oil. Repeat with a further 150 ml (5 fl oz) of grapeseed oil and the remaining mince.

Using the same pan, add the remaining 100 ml (3½ fl oz) of grapeseed oil and allow the pan to reach a light haze over a medium heat. Add the vegetables to the hot pan and sweat for approximately 15 minutes, seasoning well with a healthy pinch of salt. Cook until the vegetables have fully softened, the moisture has left the pan and the oil begins to fry the vegetables again, then add the white wine to deglaze the pan and use a spoon to lift off the sediment that has stuck to the base of the pan Add the tomato paste and cook out for 3 minutes.

Return the mince to the pan along with the tinned tomatoes, bay leaf, nutmeg and water. Bring to a boil, then turn down to a simmer so it is bubbling gently and cook out for 2½ hours until the tuna is tender. You are looking for the sauce to have thickened and reduced well, the colour to have deepened to a dark red and the oil to be starting to emerge on the surface. The finished bolognese can be stored in either the refrigerator or freezer.

FISH CONSOMMÉ

There are more modern ways of producing a fish consommé with different filtration and clarification methods, however clarifying a stock with a raft is a skill I'm glad I acquired. For this recipe we selected the bones from coral trout due to its high gelatine content and refined, sweet finish. This consommé freezes very well.

MAKES 2 LITRES (68 FL OZ/8 CUPS)

STOCK

4 fish frames
4 fish heads, gills and eyes removed
8 fish collars
200 g (7 oz) salted butter
250 g (9 oz) French shallots, peeled and diced
200 g (7 oz) celery stalks, diced
75 g (2¾ oz) whole garlic cloves
½ bunch of thyme
2 rosemary sprigs
300 ml (10 fl oz) champagne vinegar
4 litres (135 fl oz/16 cups) water
4 litres (135 fl oz/16 cups) dashi
200 g (7 oz) dried shiitake mushrooms
100 g (3½ oz) dried fish roe or dried anchovy
4 corn cobs, husks stripped
2 kg (4 lb 6 oz) mussels in shell
1 dozen oysters, shucked and juices reserved
25 g (1 oz) dried kombu
salt, to taste

RAFT

500 g (1 lb 2 oz) white fish
½ carrot, coarsely cut
1 celery stalk, coarsely cut
½ brown onion, coarsely cut
10 egg whites
pinch of salt

Preheat an oven to 190°C (375°F).

To make the fish stock, start by spreading the fish frames, heads and collars evenly across baking trays. Place trays in the oven and bake until completely golden brown all over. The heads and collars will take a little more time as there will be considerably more meat on these parts. Turn them as they cook to colour evenly.

Set the roasted bones aside until required.

In a large, heavy-based saucepan, melt the butter over a medium heat until it begins to bubble. Add the shallots, celery, garlic cloves and herbs and cook until tender, about 20 minutes. Add the vinegar to the pan and reduce down to a glaze, about another 10 minutes. Add the water, dashi, shiitake mushrooms, dried roe, corn cobs and the roasted bones. Bring the stock to a boil, then reduce to a simmer over a medium heat for 1½ hours.

Add the mussels, oysters and kombu to the simmered stock, bring it up to boil again and then remove from the heat and allow to steep for 1 hour. Strain the stock into a clean container and chill overnight.

For the raft, blend the white fish in a food processor to a coarse paste. Remove to chill in the refrigerator then add the vegetables to the same food processor and blend to a coarse pulp. Mix the blended fish and vegetables together and set aside.

In a large mixing bowl, whisk the egg whites to soft peaks. Fold into the blended fish and vegetables, then season this raft mixture with a good pinch of salt.

Decant the fish stock into a large saucepan. Whisk the egg raft into the cold stock and then place it over a low heat. As the stock comes up to a simmer, occasionally stir the raft gently to stop it sticking to the base of the pan. The objective is for the raft to set across the top of the stock and draw all the impurities out of the stock, leaving behind a beautiful clear consommé. Once the stock reaches a low simmer, cook it out for another hour, which will reduce the stock by a third.

Place a fine sieve lined with cheesecloth or muslin over a large bowl and pass the stock through carefully, ensuring the raft is left behind in the pot and any impurities are held in the sieve. This is easily done with the use of a ladle as opposed to tipping the whole stock into the sieve. Season the finished consommé. If you notice any oil or impurities on the top of the stock, remove with a paper towel placed on the surface.

The consommé is best served as a soup or can be set with gelatine for a unique textural addition to raw fish or vegetable dishes.

FISH JUS

While the Fish Gravy (see page 249) is thicker and a little more aligned with a gravy roll, this jus has a far more refined finish that will act as the mother sauce for a whole list of other conversions you wish to make. From diane to mushroom to pepper sauce, nothing is off limits.

MAKES ABOUT 225 ML (7½ FL OZ)

- 2 kg (4 lb 6 oz) clean white fish bones and trim (from the butchery of a whole fish)
- 10 French shallots, finely sliced
- 6 garlic cloves, finely sliced
- 300 ml (10 fl oz) white wine
- 200 ml (7 fl oz) white-wine vinegar
- 5 fresh bay leaves
- 20 g (¾ oz) juniper berries
- 20 g (¾ oz) whole black peppercorns
- 20 g (¾ oz) whole white peppercorns
- 1 bunch of thyme
- 80 ml (2½ fl oz/⅓ cup) dark soy sauce
- 600 ml (20½ fl oz) brown fish stock

Place the fish bones and trim in a wide, shallow frying pan that will fit them in one layer and brown over a medium-high heat for at least 15 minutes. The aim is to scrape up the sediment that settles on the base as it forms, allowing all the fat to render and the trim to crisp up. When very brown, tip everything into a colander over a bowl and allow any fat to drain, reserving the fat for later.

Return the solids to the same pan, add the shallots and garlic and cook for 10 minutes, or until lightly coloured and starting to smell sweet. Add the wine, vinegar, bay leaves, juniper berries, peppercorns and thyme and cook, stirring often, for 10 minutes, or until reduced to an almost glaze-like consistency. Add the soy sauce and stock and bring to a boil, then reduce the heat and simmer very gently, turning the fish occasionally, for 20 minutes, or until thickened and reduced.

Strain through a sieve, pressing hard on the solids, then strain again through a second clean sieve into a fresh pot and leave to rest in a warm spot so the fat separates from the sauce. Pass the warm reserved fat through a very fine sieve, then pour over the sauce. Warm the sauce over a low heat, just to heat it through without boiling. The jus is now ready to be spooned over or alongside grilled fish or vegetables. It can also be cooled and then refrigerated or frozen in a clean plastic container.

RENDERED FISH FAT

There are a huge number of fish species that carry a substantial amount of visceral fat within the cavities, so keep your eye out when gutting a fish. This rendered fat is fantastic for anywhere oil would be used, from emulsion sauces, roasting potatoes or fish, or even in making chocolate.

To render fish fat, place it in a saucepan and gently melt over a low heat; it should take about 10 minutes for it to liquefy. Strain through a fine-mesh sieve to remove any impurities. Pour into airtight containers and cool, then store in the fridge for up to 2 weeks, or in the freezer for up to 2 months.

FISH GRAVY

This fish gravy is thick, rich and glossy and a brilliant alternative to that typical 'squeeze of lemon'.

MAKES ABOUT 250 ML (8½ FL OZ/1 CUP)

2 kg (4 lb 6 oz) fish frames, including heads, fins and cartilage (gills and organs removed)
100 g (3½ oz) ghee
6 large brown onions, finely sliced
6 garlic cloves, sliced
1 fresh bay leaf
15 thyme sprigs
200 g (7 oz) fresh fish skin
300 ml (10 fl oz) white wine
2½ tablespoons sherry vinegar
1 teaspoon Vegemite
1 teaspoon dark soy sauce
750 ml (25½ fl oz/3 cups) brown fish stock
sea salt flakes and freshly cracked black pepper

Preheat an oven to 200°C (390°F). Lay out the fish frames evenly in one layer on a baking tray. Roast until well caramelised, turning every 10 minutes over the course of 30–35 minutes.

In a large pan, heat the ghee over a high heat until a haze comes over the pan, then add the onions and garlic. Stir so they are coated in the hot ghee and begin to soften evenly, then reduce the heat to medium and keep stirring until they begin to toast lightly around the edges, about 10–15 minutes. Add the bay leaf, thyme, fresh fish skin and roasted fish bones to the pan and cook for a further 10 minutes until the onion has begun to caramelise, then deglaze the pan with the wine and vinegar. Simmer over a medium heat until reduced to a syrup, approximately 15 minutes. Stir in the Vegemite and dark soy followed by the fish stock, then bring to a simmer and cook for 20–25 minutes, or until reduced by half.

Transfer batches of the contents from the pan to a food processor and pulse to a thickish, coarse sauce. Pass each batch through a fine sieve, discarding the solids, then taste and adjust the seasoning. At this point, the sauce may need further reduction.

To check the viscosity of the reduced sauce, spoon some onto a dinner plate to see how thick it is when it cools. If it is too thin, return to a medium heat and simmer for a further 5 minutes, or until the gravy is thick enough to lightly coat the back of a spoon. It is now ready to serve, or store the gravy in an airtight container in the fridge for up to 1 week or in the freezer for up to 1 month.

CURED FISH FAT

To say that fish fat is a linchpin throughout this craft section of the book is probably an understatement. Fish fat has been a profound discovery for me in further advancing and refining these charcuterie-style items.

The fat used in this recipe is cut from a Murray cod. We have found that this type of fat is predominantly found within the cavity of aquaculture species that are fed more than a wild fish would be, or well-fed wild fish in peak condition.

Being acutely aware of and looking out for this part of a fish will give you the best chance of accumulating and salting enough of it down to start putting it to work.

MAKES 1 KG (2 LB 3 OZ)

10 g (¼ oz) whole black peppercorns
5 g (⅛ oz) whole fennel seeds
90 g (3 oz) table salt
30 g (1 oz) caster (superfine) sugar
2 tablespoons fresh rosemary leaves, finely chopped
1 kg (2 lb 3 oz) fresh fish fat, trimmed

Toast the black pepper and fennel seeds separately and grind coarsely in a mortar and pestle. Sift the spices, keeping the coarse seeds and setting the powder aside for another application.

Combine the spices with the salt, sugar and rosemary. Rub the cure evenly across the fat until completely covered. Place the fat into a sterilised container and refrigerate in the cure for 5 days. After this time has passed, remove any excess liquid from the container and brush off the residual cure from the fat.

If you intend to utilise this fat for charcuterie-based applications, store in a sealed container in the freezer until needed. If using for slicing and applications that extend to immediate consumption, either wrap the fat in muslin (cheesecloth) and hang on hooks in a fan-forced refrigerator until it has dried and is firm to touch, or alternatively place the cured fat on a wire rack and allow to dry in the refrigerator for a minimum of 7 days to reach a texture that is simple to slice.

FISH-EYE ICE CREAM

When you eat 'normal' ice cream made with eggs, do you ever ask yourself, does this ice cream taste like chicken? This ice cream is in absolutely no way fishy in its taste or texture. It is to me a revelation that the vitreous humour of a fish can behave in a similar way to an egg in making ice cream.

The fish eyes must be incredibly fresh and from a known supplier, and arrive to you whole and intact on the fish. This will ensure the sanitary condition of the eye.

MAKES 1 LITRE (34 FL OZ/4 CUPS)

- 4–6 large fish eyes (or more smaller eyes) to yield 4 g (⅛ oz) of vitreous humour
- 425 g (15 oz) full-cream (whole) milk
- 12 g (¼ oz) tapioca flour
- 200 g (7 oz) overcooked steamed white rice
- 730 g (1 lb 10 oz) natural yoghurt
- 5 g (⅛ oz) fine salt
- 110 g (4 oz) dextrose
- 60 g (2 oz) caster (superfine) sugar

Start by removing the eyes and setting on a board. Using a sharp knife, make a cut in the back of an eye. The liquid that will come out first is the aqueous humour; discard this. Carefully pull out the firm pupil of the eye and attached to that will be a clear gel-like sac which is the vitreous humour. This can simply be snipped off the pupil and set aside for the ice cream. The remainder of the eye can be used in another application. Remove the vitreous humours from the remaining eyes until you have 4 g (⅛ oz).

In a saucepan over a medium heat, whisk together the milk, tapioca flour and vitreous humours. Bring the mixture to a boil, then reduce heat to medium–low and continue to whisk for approximately 10 minutes until the mixture thickens and the consistency becomes very viscous, like a creme pâtissière or white sauce.

Place the remaining ingredients in a large bowl and pour the thickened milk mixture on top.

In small batches, blend the custard in a jug blender for 5 minutes per batch. Once all blended, strain through a fine sieve into a bowl, then sit the custard over a second bowl of ice water until completely cold.

This mix can now either be churned in a commercial or domestic ice-cream machine, or ideally in a Pacojet. Follow the instructions of your chosen equipment.

Serve the ice cream with your favourite condiments. A multitude of flavours can be added to this recipe to suit any palate.

CHRISTMAS PUDDING

This is a conventional Christmas pudding, full of everything you would expect a good pud to have, with one special addition. One year I had some cured fish fat in the kitchen and thought, well, if we put an animal's suet in the mix, why not fish? Merry Fishmas.

You will need a 1.7-litre (57 fl oz) pudding basin with a lid.

MAKES 1 LARGE PUDDING

150 g (5½ oz) sultanas
175 g (6 oz) raisins
175 g (6 oz) prunes, chopped
310 g (11 oz) dried currants
75 g (2¾ oz) mixed peel, diced
125 ml (4 fl oz/½ cup) brandy
125 ml (4 fl oz/½ cup) rum
125 ml (4 fl oz/½ cup) Guinness
200 g (7 oz) Cured Fish Fat (page 250), grated from frozen
120 g (4½ oz) self-raising flour
500 g (1 lb 2 oz) dark brown sugar
320 g (11½ oz/4 cups) fresh finely ground breadcrumbs
1½ teaspoons mixed spice
¾ teaspoon freshly grated nutmeg
¾ teaspoon ground cinnamon
zest of 1 orange
juice and zest of 1 lemon
2 green apples, peeled and grated
100 g (3½ oz) toasted flaked almonds
175 g (6 oz) toasted walnuts, chopped coarsely
300 g (10½ oz) treacle
4 whole eggs

Add all the dried fruits together with the alcohol. Leave to stand for a minimum of 24 hours but up to a week to develop in flavour.

When the fruits have had their steeping time, put a pot of water on to boil to come halfway up your pudding basin or heat some water in a conventional steamer, and butter your pudding basin and the lid.

In a large mixing bowl, combine all the remaining ingredients with the macerated fruits by simply stirring it together with a rubber spatula. Scrape the mixture into the prepared pudding basin, press the mixture down and put on the lid. Crimp a square of aluminium foil over the basin lid so it is watertight.

Put the basin in the pot of boiling water or in the top of a lidded steamer and steam for 5 hours, checking every now and again that the water hasn't completely evaporated. When it's had 5 hours, remove the basin from the steamer and set aside to cool slightly (or remove from the basin and wrap in baking paper and foil once cooled and store in a cool, dry place until Christmas).

To serve, place a plate on top of the basin, turn it upside down and unmould the pudding. Serve with plenty of warm custard and ice cream, or even fish-eye ice cream (see page 253)!

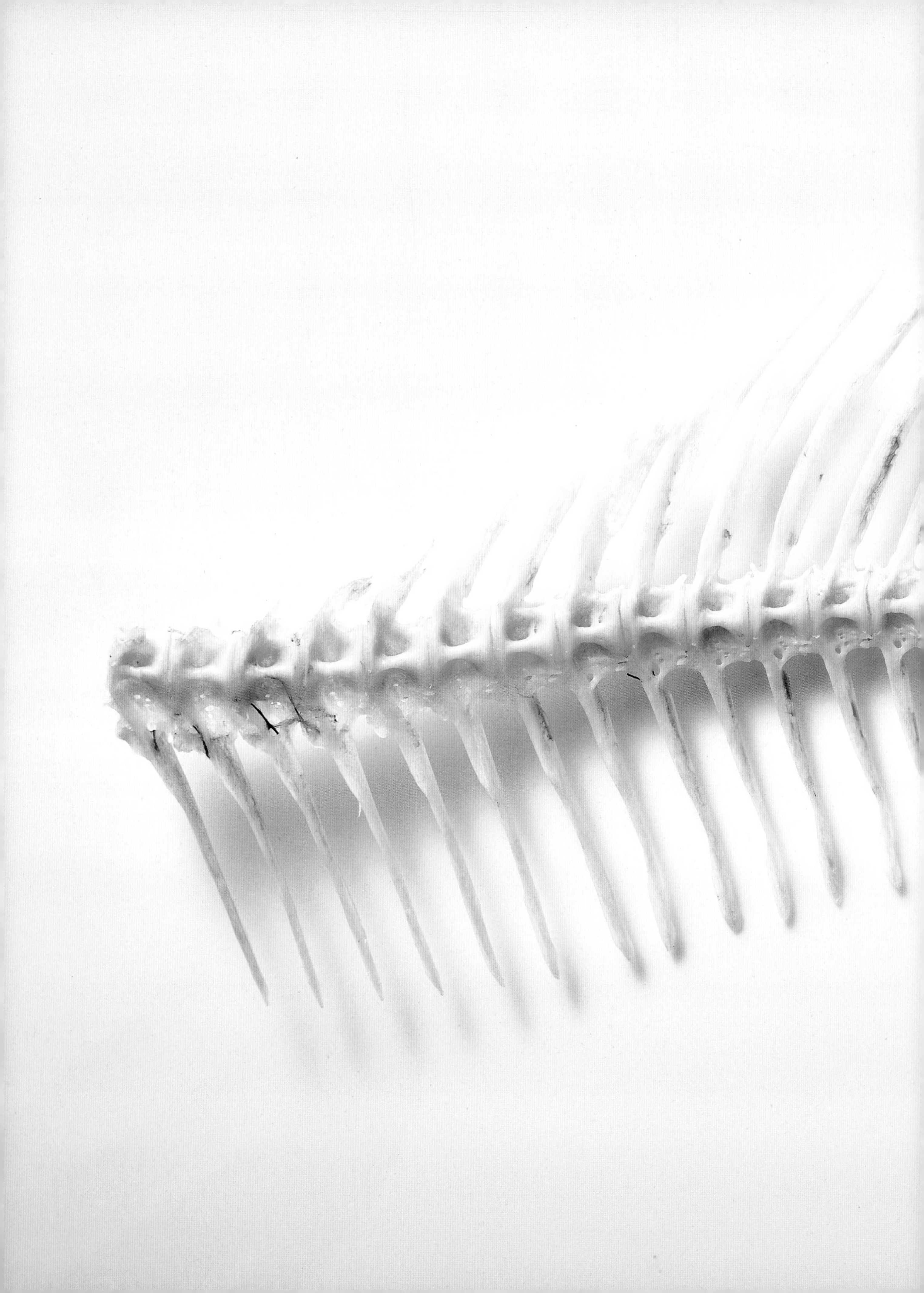

ARTISANAL GOODS

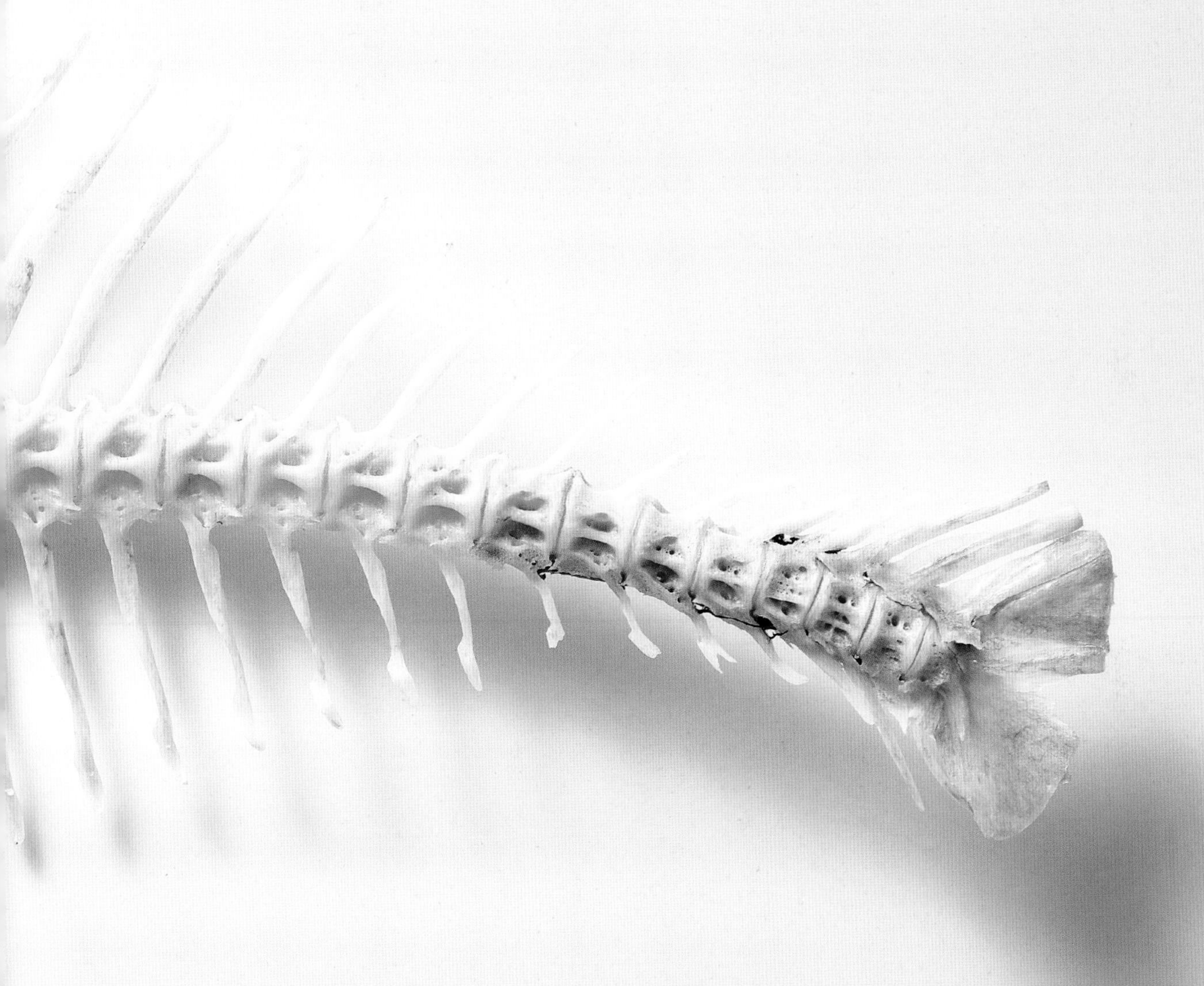

FISH-FAT COCKTAILS

BY MATT WHILEY AT RE BAR

Fish sundries over the years have been underused, especially in the bar world. But now, with the genius that Josh is showing in his restaurants and books, people like me have been afforded the opportunity to learn, taking these ingredients and using them to create special flavours. Using every part of all ingredients is really important for the future of food not just here in Australia but globally, and Josh is at the forefront of that.

We utilise fish fat in two ways. Josh and the team make an incredible cod fat caramel (featured in *The Whole Fish Cookbook*), so we put 250 g (9 oz) of this in a cryovac bag along with 1 litre (34 fl oz/4 cups) of spirit (we usually use vodka). The bag is sealed and put in a water bath at 60°C (140°F) to melt all the sugars and fat. After an hour or so, this goes into the freezer overnight. Next morning, we finely strain the liqueur through coffee filter papers and the result is an incredible salted caramel–flavoured spirit.

We also use pure fish fat in the same way as above to add viscosity and texture to drinks. Repeat the same steps as the cod fat caramel spirit but with 175 g (6 oz) of pure fish fat.

FISH-FAT SOAP

BY GRACE OF THE SOAPSTRESS

As a little girl, I remember watching my nonna rubbing a chunky bar of soap against a zigzag washboard, the bubbles doubling and tripling in size until they cascaded down her arm. I knew the soap was made from the fat of the pig they slaughtered every year, but I always wondered: how could fat clean anything?

When I grew up, had children of my own and embarked on the 'as natural as possible' route, I visited my nonna to ask if she would teach me how to make soap for my family.

Amazingly, my nonna, who couldn't read or write, made beautiful soap. She didn't have a recipe or a scientific calculator to ensure her soap wasn't 'lye heavy'. Instead, she would float an egg in a solution of water and lye. The egg served as a gauge that would reveal if there was enough or too much lye in the water to turn the fat into soap.

Armed with Google, I set off to explain the chemical reaction I had just witnessed.

The process of making soap is called saponification, an exothermic (heat-creating) chemical reaction that occurs between fats or oils and a base. What is left behind is soap and glycerine. I found this intriguing and immediately started experimenting with various fats and oils, each time discovering their different qualities and how they changed the finished soap.

Being a butcher, I naturally started with the fats I had on hand, learning how to render lard from pigs and tallow from cows. I read books from soapmakers abroad about how they would use what was local to them, including tallow from bears, deer and ostrich!

Nothing was off limits here – or so I thought. When Josh approached me to formulate a soap using the fat rendered from Murray cod, I admit to feeling very intimidated. I mean, wouldn't the soap smell fishy? And indeed, when I made a trial batch using only this fat, it most definitely was – no amount of essential oils was going to help me there! It was also really soft and didn't feel that amazing, to be honest.

Not discouraged, I made a second batch, this time, using 50 per cent Murray cod fat blended with other oils that would promote hardness, lather and moisturising qualities. Not bad, I thought to myself. I was close, but it still wasn't perfect. I played around with different percentages of Murray cod fat until I hit nirvana. Now, I don't say nirvana lightly. I'm a soap connoisseur – honestly, more like a soap snob. And this blew my socks off!

As I rolled the soap in the palms of my hands, the rich, buttery lather built up and my senses were immediately heightened by the release of the beautiful blend of essential oils, and the soap felt silky and rich. As I rinsed, my hands felt as though I had applied hand lotion. I knew the texture of the soap was directly linked to the high linolenic acids from the fish oil. It was perfect, and I have no doubt that we will be seeing a lot more fish fat soap in the years to come.

FISH-FAT CANDLES

BY HUNTER CANDLES

Candle making is a somewhat mysterious craft. As a chandler, I can't help but obsess over the wax, the wick size, how the candle burns, how it sets, how long it lasts, how well the scent throws ... And that's just the beginning. I've been honing my craft for fifteen years now as Hunter Candles, yet I am still discovering, learning and experimenting each and every day, which only further fuels my addiction.

When Josh, Julie and Ben bustled into our store one fresh Sydney morning, I knew exactly who they were – their reputations precede them. But I could never have guessed the opportunity they were about to present to me: to create a candle from their spare Murray cod fat. The nose in me piqued as I imagined what pure blubber from a Murray cod might smell like – bad, I presumed. Yet the more I mulled it over, a fabulously fatty, muddy, seawater burn began swimming around my mind. Naturally I jumped at the opportunity and began impatiently awaiting the arrival of my blubber.

While it might sound subversive, the concept of fish-fat candles has been around since the eighteenth century. Wax from sperm whales was used by chandlers as it burned cleanly and didn't produce an unpleasant aroma like some tallow candles. While sperm whale wax is no longer used, the concept itself still serves as inspiration.

On arrival, the 'fishy' smell intrigued me – without hesitation, I threw my nose straight into the bag of blubber. It's a scent that's hard to pin down, but to me it smelled soft and salty, like a wrist without perfume. And somehow not like fish at all.

I dived headfirst into experimentation, meticulously weighing, smelling, mixing and stirring, tweaking a little each time. Wax and fat are fabulous friends, and when heated they blend together beautifully. The Murray cod fat glistened. It has a thinner consistency than soy wax, so it took longer to cure, blend and harden. But once it did, I looked for crystallisation, aesthetic inconsistencies and density. The colour was creamy, the wax fat hard. From here it was a case of trial and error, the first of which saw rogue chunks in the candle catch the flame and crackle the fat. In candle making that uses oils, this is known as the flash point.

I liked the melty fat smell and ended up using the rendered fish fat, which was velvety and had a low melt point, marrying with the wax perfectly. The end result was creamy to burn, and the natural scent had a subtly sweet note, so we focused on additional scents that would complement it, something intriguing and elegant that would melt away any hesitation that may come with burning a candle made from discarded blubber.

We looked to where you find Murray cod, their natural habitat. They mostly swim in rocky streams and billabongs framed with fallen red gums and moss-covered rocks. From there we fell into green botanicals, muddy mossy notes, a touch of pink peppercorn and a woody cedar base. Ultimately raw, earthy, yet still fresh in a unique way. We've called it Murray Cod Pepper Rocks.

Lighting our candle for the first time, a herbaceous, mossy breeze with an unmistakable earthy base and slight sprinkling of savoury pepper envelops the space. It's a scent that can absolutely hold its own, anywhere, anytime. I am eternally grateful to Josh, Julie and Ben for bringing this type of craft into the conversation. It has recalibrated the way I think about how candles are made, and about what other readily available possibilities may be out there in less common mediums. It's only sensible that we keep pushing the boundaries of what's possible when it comes to sustainability.

HUNTER

FISH-BONE CERAMICS

BY SAM GORDON

As a third-generation potter, clay has always been a constant in my life. School holidays from as early as I can remember were spent with pottery, from making a mess to weighing clay for the potters.

Supplying the hospitality industry with venue-specific plates has always been my passion. The opportunity to use bones that would otherwise be a waste product in any other kitchen (even after stocks have been made) in the glaze of the crockery is extremely rewarding.

The use of bones in either the body of clay (bone china) or the glaze is not new and was first done in the 1750s. The bones from Fish Butchery are first boiled to remove any tissue, then I fire the bone alongside my hand-thrown crockery in a vessel called a bisque, the temperature gets up to 1000°C (2264°F) so the bones calcify. This firing takes 8 hours, then once the bones are cool to touch they are easily crushed into a powder. I add this powder to either my white or clear glaze recipe, which is essentially a liquid glass comprising of silica, aluminium oxide and fluxes that melt at temperature. I require 1240°C as the top temperature.

The beauty of adding a percentage of bone ash to the glaze is that it makes a glaze translucent and helps it to melt into the body, plus we are using a product that would otherwise be waste. The use of the bones in ceramics truly closes the loop between a restaurant, fish market and potter. But this is only the beginning – we have much more to explore with bone ash in glazes and in a pure bone china that requires 40 per cent bone ash.

INDEX

FISH BUTCHERY

TAKE ONE FISH

ACKNOWLEDGEMENTS

Once again, having the opportunity to express my thoughts, views, and practices with regards to fish is something for which I am incredibly grateful. And the only reason I have the freedom to explore and develop techniques – both old and new – is the incredible people around me.

First and foremost, to my beautiful wife Julie, without whom Saint Peter, and all the businesses we have created together, would have never existed (nor survived). Your endless trust, courage and tenacity continues to inspire me every day.

Rebecca Lara, you are the backbone of Fish Butchery. Your enduring work ethic and tremendous technique has motivated and inspired those around you for all the years we have worked together.

To Ben Torrance, you are the one who got the wheels turning and allowed me the freedom to be at my very best. Not only are you an incredibly gifted chef and leader but a wonderful friend and driving force behind the work we do.

Bosley McGee, you are as rare as a needle in a haystack. What you bring to Fish Butchery and our entire team is invaluable. Thank you for your patience, technical prowess and good humour throughout the years we have worked together and help in making this book what it is.

Chris Karvellas, thank you for steering the ship through any conditions, for loving the team and our businesses as if they were your own and for always being a beacon of possibility and positivity.

To Simon, Daniel, Rob, Lucy, Michael, Roxy, Julie, and the entire team at Hardie Grant, past and present. Thank you for your belief in me and for your focus and attention in bringing this beautiful book to life.

Further thanks to everyone who contributed to this book – from the brilliant craftsmanship of Claudio Morales to the very thoughtful words written by Luke Buchholz, Darren O'Rourke, Tony Wearne, Matt Whiley, Vianney Hunter, Sam Gordon and Grace Steven, not to mention the masterfully designed illustrations by the extraordinary Reg Mombassa.

This book would be nothing without you all and I am forever grateful. Thank you.

– Josh

Published in 2023 by Hardie Grant Books, an imprint of Hardie Grant Publishing

Hardie Grant Books (Melbourne)
Building 1, 658 Church Street
Richmond, Victoria 3121

Hardie Grant Books (London)
5th & 6th Floors
52–54 Southwark Street
London SE1 1UN

hardiegrantbooks.com

A catalogue record for this book is available from the National Library of Australia

Fish Butchery
ISBN 978 1 74379919 2

10 9 8 7 6 5 4 3 2

Publisher: Michael Harry
Project Editor/Editor: Simon Davis
Copy Editor: Simone Ford
Design Manager: Kristin Thomas
Designer: Daniel New
Photographer: Rob Palmer
Stylist: Lucy Tweed
Production Manager: Todd Rechner

Colour reproduction by Splitting Image Colour Studio
Printed in China by Leo Paper Products LTD.

The paper this book is printed on is from FSC®-certified forests and other sources. FSC® promotes environmentally responsible, socially beneficial and economically viable management of the world's forests.

Hardie Grant acknowledges the Traditional Owners of the country on which we work, the Wurundjeri people of the Kulin nation and the Gadigal people of the Eora nation, and recognises their continuing connection to the land, waters and culture. We pay our respects to their Elders past, present and emerging.